Writing and Designing
Manuals and Warnings

Writing and Designing Manuals and Warnings

Fifth Edition

Patricia A. Robinson

CRC Press
Taylor & Francis Group
Boca Raton London New York

CRC Press is an imprint of the
Taylor & Francis Group, an **informa** business

First published in paperback 2024

First published 2020 by CRC Press
2385 NW Executive Center Drive, Suite 320, Boca Raton FL 33431

and by CRC Press
4 Park Square, Milton Park, Abingdon, Oxon, OX14 4RN

CRC Press is an imprint of Taylor & Francis Group, LLC

© 2020, 2024 Taylor & Francis Group, LLC

Library of Congress Cataloging-in-Publication Data

Names: Robinson, Patricia A., 1948– author.
Title: Writing and designing manuals and warnings / authored by
Patricia A. Robinson.
Description: Fifth edition. | Boca Raton : CRC Press, 2020. | Includes
bibliographical references and index.
Identifiers: LCCN 2019036123 (print) | LCCN 2019036124 (ebook) |
ISBN 9780367111090 (hardback) | ISBN 9780429025372 (ebook)
Subjects: LCSH: Technical writing.
Classification: LCC T11 .R635 2020 (print) | LCC T11 (ebook) | DDC
808.06/66—dc23
LC record available at https://lccn.loc.gov/2019036123
LC ebook record available at https://lccn.loc.gov/2019036124

ISBN: 978-0-367-11109-0 (hbk)
ISBN: 978-1-03-292243-0 (pbk)
ISBN: 978-0-429-02537-2 (ebk)

DOI: 10.1201/9780429025372

Visit the Taylor & Francis Web site at
www.taylorandfrancis.com

and the CRC Press Web site at
www.crcpress.com

Dedication

This book is dedicated to all my students, in classrooms and seminars, in formal and informal settings. I've learned more from you than I can say.

Contents

PART I   Product Safety in the 21st Century

PART II   The Making of a Manual
(Or More Than One!)

PART III   *You Have Been Warned*

PART IV   Making It Work in the Real World

Preface to the Fifth Edition

The world is different from how it looked even a decade ago, and the pace of change just keeps accelerating. When the last edition of this book was written, the iPhone had just been invented. Now smartphones are everywhere and doing things unimaginable just a few years back. You can use your phone to track your escape-artist dog, turn the heat up in your house as you commute home from the office, accept a digital payment, and board a plane. And that's just for starters. Technology is changing the way we do business, the way we communicate with each other, and the way we learn.

More companies than ever are doing business internationally, but even domestically, the U.S. population is increasingly diverse. About 20% of the population speaks a language other than English at home, and that percentage is expected to remain steady for at least the next decade. Providing product users with instructions and safety information becomes even more of a challenge when customers span an ever-widening wide spectrum of languages and cultural variation.

Since the last edition, products liability exposure has become even more of a concern for manufacturers, and allegations of failure to warn have become a leading basis for lawsuits in the U.S. At the same time, while litigation is less prevalent internationally than here at home, new international standards and country-specific legislation have recognized the critical role played by instructions and warnings.

This book, like its predecessors, is intended to help those in the trenches—the technical writers, graphic artists, engineers, and others who are charged with producing product documentation. While the content is informed by current research, it also reflects the real-world knowledge of hundreds of practitioners. Research can guide us, but there is no substitute for experience when it comes to applying that guidance when the realities of budgets and market-driven timelines make trade-offs inevitable. Whether you are brand-new on the job or have labored for years, this book will have something to help you.

Some things don't change, but many do. This edition, while preserving the basic guidelines for developing manuals and warnings presented in the previous edition, offers new material as well, including an expanded section on hazard analysis. As litigation involving warnings and instructions increases and products connect in the Internet of Things, it becomes even more important to have a systematic way to evaluate and respond to product hazards. And while virtually all products need instructions and safety information, this information may not always be presented in a paper manual. This edition explores into how emerging technology is changing the world of documentation, from videos to virtual reality and all points in between.

As always, I hope you find it a helpful resource to navigate the challenges of product documentation in the 21st century.

Patricia A. Robinson

Acknowledgments

Many individuals have contributed to this and previous editions of this book by sharing their expertise and their materials—too many to list here. I thank you and so do the readers of this book who will use your contributions to write better manuals and develop more effective warnings. One of the book's central motifs is to focus on *keeping the user safe*. Because of the collective wisdom you have freely shared, be assured that you have helped to do just that: you have prevented injuries and saved lives.

The materials for the book were collected in a number of ways. Some came from in-house technical writing seminars for industry, visits to service publications operations, teaching in national and regional conferences and workshops devoted to product safety and technical writing, and serving as a private consultant. Individuals and organizations who have worked with us and/or whose materials provided the subjects and examples include Acme Burgess, Inc.; American Honda Motor Company; American Optical Company; American Suzuki Motor Corporation; Atwood Mobile Products; Butler Manufacturing Company; Briggs and Stratton Power Products Group, LLC; Chrysler Motors Corporation; Clarion Safety Systems, Inc.; Clark Components International; Construction Industry Manufacturers' Association; CooperVision Surgical Systems; Country Home Products, Inc.; CPM Acquisitions Corporation; Datamax Corporation; Deere and Company; Delta International Machinery Corporation; Devilbiss Company; Doboy Packaging Machinery; Dri-Eaz Products, Inc.; E.1. DuPont De Nemours and Company; Electrolux Home Products; Epson America, Inc.; Farm Implement and Equipment Institute; Fellowes Manufacturing Company; Fisher and Paykel Appliance, Inc.; FMC Corporation; Ford Motor Company, Ford Tractor Division; Ford New Holland; Fulton Manufacturing Corporation; Gardner Denver; Gehl Company; General Electric Company and General Electric Medical Systems; General Motors Company; Gerber Products Company; Harley Davidson Motor Company; Hazelton-Rattech Inc.; Hewlett-Packard; Honeywell Inc.; Huffy Corporation; Ingersoll-Rand Company; International Business Machines; International Harvester Company; International Organization for Standardization; J. 1. Case Company; John Muir Publications; Joy Manufacturing Company; Kartridg Pak Company; Kohler Corporation; Krones Inc.; Mack Trucks-Mack International; Madison-Kipp Corporation; Mantis Corporation; Martin Engineering; National Electrical Manufacturers Association; Norden Laboratories; Ohio Medical Products; Outboard Marine Corporation; Pierce Manufacturing Company; Pittsburgh Poison Center; Prairie Agricultural Machinery Institute; Prairie Implement Manufacturers' Association (and affiliated companies of Canada); Rheem Manufacturing Company; Rosemount Inc.; Ryobi Outdoor Products, Inc.; Scotsman; Siemens Medical System; Silver Reed America, Inc.; Simplicity Manufacturing, Inc.; Sologear, LLC; Teresa Sprecher; Rosemary Stachel; Sunstrand Aviation; TRW Ross Gear Division; Taylor Instrument Company; Technicare; Thern Inc.; Versatile Corporation; Volkswagen of America; Wabco Construction and Mining Equipment; Weber-Stephen Products Company; Westinghouse Electric Corporation; World Kitchen, LLC.; and Yamaha Motor Company.

Finally, I owe a special thank you to my family and friends for putting up with me through yet another edition of this book.

Author

Patricia A. Robinson has been involved in technical writing for forty years. After earning her PhD (in a different field) she joined the faculty of the University of Wisconsin-Madison College of Engineering where she taught technical writing for thirteen years. During her tenure at the University, the technical communication program grew from a single report-writing course into a full certificate program. She served as the first Director of that program. Dr. Robinson also served as program director and primary instructor for continuing education seminars for practicing technical writers and engineers, covering everything from Effective Speaking to Managing the Technical Publications Department. An early invitation to teach an in-plant course on operator manuals for a farm implement manufacturer led to an unexpected—but greatly satisfying—career focus on instructions and warnings.

Dr. Robinson is the author of *Fundamentals of Technical Writing* and co-author of *Effective Writing Strategies for Engineers and Scientists*. She has also published articles in various professional journals and presented papers at professional conferences. She is a regular speaker at product liability seminars on the role of instructions and warnings. Dr. Robinson is a Full Member of the Human Factors and Ergonomics Society, a member of IEEE, and a member of the ANSI Z535.6 subcommittee.

As owner of Coronado Consulting Services, LLC, Dr. Robinson consults widely for industry both in the United States and internationally, and occasionally provides expert testimony related to instructions and warnings. Dr. Robinson presently makes her home in a small southern Arizona community, where horses and cattle considerably outnumber humans. She can be reached at pat@coronadoconsulting.com.

Part I

Product Safety in the 21st Century

1 The Changing Landscape

OVERVIEW

The first edition of this book was published in 1984. To put that in context, the first personal computer (desktop) had been released only about ten years earlier. Illustrations for manuals were drawn by humans wielding technical pens. Thirty-five years later, we live in a much different world. Yet, most tangible products (as opposed to software) are still accompanied by a manual—of some kind. Today that "manual" may be a YouTube video or a QR code on the packaging that takes you to a website with animations and .pdfs of instruction sheets. Sometimes the instructions for using a product are built right into the product, especially if it has a screen as an interface. But quite often you still get a paper manual.

Still, much has changed—and the pace of change is only accelerating. Customers and companies have much higher expectations for products and manuals than they did in 1984. The definition of "product" has expanded to include not just the instructions and warnings that accompany it, but sales and marketing materials as well. Technology has altered the ways in which we interact with products and is even allowing products to interact with each other without human input. And international commerce and multinational companies are now the norm rather than the exception, bringing along proliferating international standards and regulations. The previous edition of this book devoted a single chapter to writing manuals for the global marketplace. This edition assumes your manuals will be used worldwide. And it assumes that your product will still be accompanied by a manual—in some form.

WHY DO WE STILL NEED MANUALS?

If you're reading this book, chances are you are engaged in some way with manuals. You may be a technical writer laboring to produce manuals. You may be an attorney involved in a products liability lawsuit. You may be in charge of product safety at your company. You may be a marketing director who recognizes that the manual is perceived as part of the product "package." This book focuses on writing and designing standard product manuals composed of words (most of the time) and graphics, whether the product user sees them in paper or digital form.

Manuals, in whatever form they appear, are still around because they fulfill a variety of necessary functions. Some of these are obvious, but others, while less apparent, are just as important—maybe more so. One way to look at the functions of a manual is to consider how a good manual benefits both the customer and the company—or how a bad manual hurts them.

THE CUSTOMER

The obvious function of a manual is to educate the customer to use and maintain the product properly. Manufacturers have a legal duty to tell their customers how to safely use and care for their products (more about this duty in Chapter 9). It doesn't matter whether the product is a hundred-dollar string-trimmer purchased by a home-owner or a million-dollar veneer press purchased by a wood products company. The duty is the same—though the products and users are different. As we shall see, the legal duty is to provide *information*, but the underlying need is to facilitate proper and safe *behavior*. Accomplishing these goals requires close attention both to what you say and how you say it. Essential information packaged in incomprehensible prose does little good—but the most readable manual in existence won't help if critical information is missing.

Regardless of the vast differences among them, all customers expect a manual to serve three basic purposes:

- Explain how to operate (or maintain) the product the first time they use it.
- Provide reference information they can use throughout the product's life.
- Serve as a safety resource.

These three functions are quite different, and require different organizational and writing strategies to fulfill them.

FIRST-TIME OPERATION

The user operating the product for the first time naturally needs more information and guidance than an experienced user of the product or similar product. If it's your very first time using a new product, you need to know everything about what to do—and what not to do—to operate the product properly. Particularly if the product is itself brand new (that is, innovative), the writer must be sure to provide enough information to the first-time user. That's not as easy as it sounds. In fact, one of the most difficult tasks of the technical writer is to include all the information that the first-time user needs.

Why is it so hard? Simple: because as a technical writer working for the manufacturer, you've been living and breathing the product for a matter of weeks, if not months. Through countless discussions with engineers and other product developers, you know the product like the back of your hand. If you have spent eight hours a day for several weeks immersed in developing sufficient expertise to write about the product, it's difficult to remember what it was like the first time you encountered it. You have become so familiar with it that you're apt to assume knowledge on the part of the user that he or she just doesn't have. If you misjudge how much your reader knows, the result is a confused and frustrated user.

What does the first-time user need? In general, first-time users need these categories of information:

- Intended uses and limitations of the product.
- Names and locations of parts and controls.

- Step-by-step procedures for installing, starting, operating, shutting down, maintaining, and (sometimes) storing the product.
- Safety information and warnings about hazards associated with using the product.
- Troubleshooting information and (sometimes) repair procedures.
- Contact information for the manufacturer.

Unless you have learned to look at your product as a first-time user, it's easy to forget some of them—especially the first two categories. When your own company manufactures the product, you tend to take for granted what it's used for and what the specialized terminology associated with it means. A first-time user may have a general idea of the product's purpose, but not know all the possible uses and certainly may not know its limitations. And unfamiliar terms can put up a sizeable barrier for the first-time user.

ONGOING REFERENCE

A second customer-related function of a manual is for use as an ongoing reference. Many manual writers openly acknowledge this purpose by advising the reader to "read and save these instructions." For consumer products, this admonition often results in a kitchen drawer crammed full of an assortment of manuals and instruction sheets. For industrial products, the equivalent is shelf upon shelf of three-ring binders lined up in the purchasing agent's office. Either way, the user has a source to look up specific information that is not needed on a daily basis: How do you change the filter for the refrigerator water dispenser? What's the procedure for lubricating the bearings on the outfeed conveyor? What's the tolerance for blade wear on the slitter machine? The answers to all these questions (and many more) make up the reference function of a manual.

Reference material is every bit as vital to users as first-time operation instructions, but it is not nearly as easy to organize and write. Writers must not only understand and be able to explain an astonishing variety of processes and procedures, but they must also anticipate the user's need for information and organize the information for easy retrieval. If the information is too difficult to locate, users are apt to become frustrated and give up looking for it. They will then try to get along without the needed information—either by trying to figure out what to do on their own or by contacting the manufacturer for technical assistance. Either option can be costly.

PRODUCT SAFETY RESOURCE

Another function of a manual for customers is to provide comprehensive information on using a product safely. As we shall see, on-product warning labels are usually reserved for the most severe hazards, and are limited in the amount of information they can convey before they lose their effectiveness. The manual can expand on label information as well as give additional safety-related information about the product. For example, the label on the product may warn the user to "keep hands away from moving parts." While that message is important, it leaves some questions unanswered:

How far away should I keep my hands? How close is too close? Where *should* my hands be when I'm operating the product? How do I clear a jam in those moving parts? How do I lubricate them? The manual can answer all those questions in detail.

The manual can do much more than simply expand on product labels. Not every hazard associated with a product needs to have a corresponding warning label. In fact, the more labels plastered all over the product, the less effect each one has. The effect is similar to the way that putting "New! Improved!" on the dish detergent label will catch the consumer's eye—unless every other brand has the very same words on their label. In that case, no one stands out more than the others. For that reason, on-product warning labels should be limited to the most serious hazards. But what about the other hazards? How do users learn about those?

Manuals offer an opportunity to expand on safety information. In some cases, all that is needed is a fuller explanation of information on the label. But often the user—especially the first-time user—needs much more. In some circumstances, the manufacturer may provide a separate safety manual that explains safe working procedures for a class of products—for example, printing presses or air compressors. Other times, the manufacturer may include a safety section at the front of the manual that includes general safety information as well as warnings specific to the product. In all cases, the manual is expected to include at a minimum, warnings relating to the product it covers.

THE COMPANY

A well-designed manual (or manuals) clearly benefits the customer by providing needed information in a convenient and accessible package, but manuals also benefit companies that manufacture products. One of the changes in the field over the last 35 years is the recognition that product documentation does not merely add cost to the manufacture of products—it actually adds value. The value-added aspect of a good manual becomes increasingly clear as companies move more and more into global marketplaces. Good product documentation is a requirement for European Union compliance and for ISO certification. For ISO 9000 quality assurance, for example, all but two of the 20 major system elements explicitly mention documentation as vital, and the other two elements imply it. Whether in the United States or abroad, good documentation provides tangible benefits for the company as well as the customer.

REQUIRED DOCUMENTATION

While providing operational and reference information that users need, manuals also provide historical information companies need. Products do not remain static; manufacturers continuously seek to improve them. Whether in response to customer feedback or marketplace competition, the engineers are always tweaking this and altering that. Depending on the product, this constant change process can be very rapid or relatively slow. Either way, the product itself is always a moving target.

Ongoing evolution of product design is the source of one of technical writers' biggest frustrations: just when you think you have the manual draft complete, the engineers change something and the rewriting begins. It's also a source of frustration for users, because rarely will a company wait for the documentation to catch up before shipping the product out the door. As a consequence, the user is often looking at a manual that is a little out of date. For a first-time user who is not already familiar with the product, the difference between the product in hand and the description of the product in the manual can create a considerable obstacle to understanding.

Even so, manuals become the historical record of a product's evolution. So what? Other than to satisfy an academic interest, why would anyone need such a thing? The author of a book on writing manuals may need a garage full of old manuals to use for illustration, but who else would? In fact, you and your company need that historical record as well. Knowing exactly what the product and the documentation included at a particular time is important for two reasons:

- Some products have a long, useful life and may be resold multiple times.
- Product documentation may be important in a products liability lawsuit.

Both reasons create an incentive for accurate record keeping.

Many products, especially large capital equipment used in manufacturing, typically last for years. Some industrial machines may remain in use literally for decades, and may be resold multiple times. Even some consumer products may have a surprisingly long useful life—fostered by the growth of the Internet! The popularity of auction sites such as eBay has given new life to many consumer products. Unfortunately, whether the product is a gravel separator or a tabletop grill, the manual does not always accompany the product to its subsequent owner. Without keeping good records, the manufacturer may not be able to assist a new owner with use of the product. Too often, design changes and the associated instructional variations are documented only in the mind of the company employee with the longest tenure. When he or she retires, that information is lost. Keeping a library of product manuals can be crucial to providing good customer service.

As we shall see, allegations of failure to warn are increasing as a cause of action in products liability lawsuits. To defend against allegations of failure to warn, manufacturers need to be able to show that the warnings and instructions that accompanied their product were in fact adequate. It's difficult to do that if you no longer have the manual. Storage of product documentation takes up valuable space and requires effort to maintain, but in the long run, a complete library of manuals may save the company a great deal of money.

PRODUCT DEVELOPMENT RESOURCE

A second use for a manual within a manufacturing company is as a product development resource. We know that products constantly evolve to meet customers' needs—but how do companies find out what those needs are? One way in which the manual can assist in this process is by pointing to aspects of the product that need improvement. Two indicators of needed design improvements are

- The presence of lots of safety warnings.
- Complex procedures for commonly required tasks.

Whenever you find that a great many safety warnings are needed for a particular aspect of the product, there's a good chance that the design could be improved. Not every hazard can be eliminated, of course, and some products are inherently somewhat dangerous to use—a table saw, for example. The tip-off to a needed design improvement is not so much the number of warnings *per se*, but rather their distribution. Whenever you find one aspect of your product (blade changing, for example) that requires an inordinate number of warnings compared to other aspects of operating and maintaining the product, look for design improvement opportunities.

Similarly, whenever accomplishing a routine or common task requires a complex procedure (relative to other procedures for the same product), you may well be able to find a design improvement that will simplify things. Often, complex procedures signal that a product has been adapted to a use for which it was not directly designed. A redesign may achieve two goals at the same time: broadening product applicability and simplifying its use.

Just as more moving parts in a machine mean more opportunities for something to break, lots of warnings and complex procedures increase the likelihood that users will fail to heed one of the warnings or fail to follow some of the instructions, resulting in the potential for injury or damage, or at the very least, frustration and poor results. Good engineering produces the optimum balance between function and usability—a product that both works well and is easy to use. The manual can help you spot areas where that balance needs to be restored.

MARKETING AND PUBLIC RELATIONS DOCUMENT

One of the little-recognized but most important functions of a manual is to serve as an ambassador for the company. Just as a diplomat for a nation conveys a message for its head of state, a user manual sends a message to the buyer of your product. We generally think of manuals as conveying information about the product, but they also convey information about the company that made the product. Sometimes the message they send is positive. A clear, user-friendly, well-organized, and attractive manual conveys that the company is competent, wants to make using their product as easy and straightforward as possible, and cares about their customers' safety and well-being. A poor manual says just the opposite—and even worse, can bias the user against the company and the product.

You're not likely to look forward to exploring the wonders of your new smart TV if the set-up instructions are so confusing you can't even figure out how to make the sound work, much less control it from your smartphone. How are you likely to view the company that made you feel that way? Or if the manual tells you that the humidistat on the commercial dehumidifier needs to be manually calibrated once a month, but fails to tell you how to do that, will you call technical assistance to find out, or just ignore the instruction and dismiss the manual as having been written by idiots or jerks?

By contrast, a good manual adds real value to the product. Never underestimate how much effective documentation can do to increase customer satisfaction and generate repeat business. Companies have come to realize that the real commodity they sell is the company's *brand*. That brand identity permeates the product, the service, the sales effort, and all the collateral materials that accompany the product—manuals, installation instructions, "quick-start guides," specification sheets, and so on. As the late advertising executive David Ogilvy put it, "You . . . have to decide what 'image' you want for your brand. Image means personality. Products, like people, have personalities, and they can make or break them in the marketplace."

Today, companies want their customers to associate the company brand with top-notch quality regardless of whether they are looking at the product, the documentation, the company's role in the community, or something else. This shift to branding presents an important opportunity to argue for putting more resources into producing superior product documentation. After all, if the manual is the company's voice; shouldn't it communicate as effectively as possible? First and foremost, public relations professionals must be good communicators. If they aren't, they're not doing their job. Manuals, for good or ill, are public-relations documents, and as such, should reflect well on the company that produced them.

WHAT'S CHANGED?

So far, the landscape doesn't look all that different from how it did in 1984. The documents may be in digital form, but the functions of a manual are pretty much the same as they always were. What has changed, however, is what we expect of products and manuals and the environment in which they are used. Four sea-changes have taken place that affect how we write and design manuals and warnings.

- Customers expect much more from products and manuals.
- The definition of "product" has expanded.
- Technology permeates everything.
- We operate in a global marketplace.

HIGHER EXPECTATIONS

In 1984 the Consumer Product Safety Commission (CPSC) and the Occupational Safety and Health Administration (OSHA) had been in existence for less than 15 years. While certain products (notably food and drugs) had been subject to safety regulations for many years, most products (and workplace safety practices) went largely unregulated before the 1970s. The rate of non-fatal workplace injuries causing lost workdays reflects the increasing emphasis on safety: in 1979, it was approximately 4 per 100 workers, but by 2016 the rate was only 1.6 per 100—a 60% decrease.[1] Products and workplaces are safer today than ever before.

SAFER PRODUCTS

Today we *expect* products to be safe. Products that just a few decades ago were acceptable—a table saw with an unguarded blade, for example—would now be considered unreasonably dangerous. Lawnmowers and industrial machines have "dead man" switches so that if the operator's hand leaves the control, the machine stops.

What does this have to do with writing product manuals? Quite a lot, as it turns out. Even though products have become safer over the years, people continue to be injured and products liability lawsuits continue to be filed—and increasingly, those lawsuits allege a failure to warn and/or instruct as a direct cause of the injury. That makes sense, given that product *designs* have become much safer, in part because of regulation and in part because of litigation.

BETTER INSTRUCTIONS AND WARNINGS

One of the not-so-obvious uses of a manual is as a key element in a products liability lawsuit. These days, when someone is accidentally (or even intention-ally!) injured or killed, a products liability lawsuit is almost certain to follow. In a products liability lawsuit, the plaintiff alleges that the product involved was "defective" in some respect, and that the alleged defect caused the injury or death. Why do attorneys do this? They want to increase the chances that their client will be able to collect adequate money damages. If an employee was severely burned in a flash fire when he used a portable halogen work lamp that ignited flammable paint vapors, suing the employer for failing to provide proper lighting won't do much good if the mom-and-pop painting contractor declares bankruptcy. On the other hand, if you can show that the lamp was defective, you can go after the lamp manufacturer, which is much more likely to have insurance to cover a money award.

Increasingly, products are alleged to be defective because of inadequate instruc-tions and warnings. Clearly, if the plaintiff is citing the instructions and warnings as defective, the manual and any on-product warning labels are going to be a cen-tral issue in the court case. Whenever a lawsuit involves instructions and warn-ings, you can bet that the manual will be read, reread, and thoroughly dissected by experts on both sides of the case. If you are used to your work as a technical writer going ignored and unsung (nobody reads manuals anyway, right?), a failure-to-warn case can come as a real shock. Your choice of words, your organizational structure, even sometimes your punctuation, will be scrutinized, analyzed, and trotted out as evidence to bolster one position or another. If there is any inconsis-tency or contradiction between statements in different parts of the manual, you can believe that mismatch will be paraded before the jury as evidence of your callous and willful disregard of the user's safety and well-being. Offhand writing deci-sions made under the pressure of a deadline can come back to haunt you when you least expect it.

This increased attention paid to product documentation is not all bad (unless you're the one sitting in the deposition, that is). Forty years ago, regrettably, tech-nical writing was often seen as a dead-end job for people who couldn't cut it else-where in the company. As products liability lawsuits began to incorporate product documentation as a potential source of defect, companies began to take manuals and warnings more seriously. A side effect was the dawning recognition that the people who produce good, solid documentation are knowledgeable professionals deserving of respect. In part as a direct result of lawsuits, the overall quality level of product manuals has risen exponentially.

INFORMATION ON DEMAND

At the same time as manuals have gotten better, the potential sources of product information have proliferated. If the manual isn't handy, just do an Internet search about how to use the product and you will probably find everything from the manufacturer's website (which presumably provides accurate information) to user forums and YouTube videos, which may (or may not) provide proper operating instructions. We have become accustomed to getting information on demand. This has major implications for writers of manuals: product users expect to be able to find the information they seek *right now* without having to fumble around flipping pages and reading through irrelevant material. Manuals that are not user-friendly are likely to be ignored.

Technical writers are expected not only to write clear, user-friendly manuals, but may be consulted (or drafted) to develop product videos, social media campaigns, or other technological ways to convey safe operating instructions. Much as desktop computers transformed manual production from a series of discrete steps completed by specialists (writers, graphic artists, photographers, layout artists, paste-up artists, etc.) to a more fluid process in which the technical writer might also create or import digital drawings, take photos with a digital camera, and produce the layout and final copy electronically using publishing software, the Internet has further transformed the job. Nowadays a technical writer is much more than a maker of manuals—he or she is expected to be skilled in communicating essential information to users through a variety of media.

DEFINITION OF PRODUCT

Twenty or 30 years ago, businesses tended to think of their product as the actual, tangible widget they manufactured. Everything else—the manual, the packaging, sales literature—was viewed as just ancillary material that added cost. No longer. Companies have come to realize that the product is not only the manufactured item, but also the complex of information that surrounds it. The European Union's Machinery Directive for example, refers specifically both to instructions and sales literature:

1.7.4. *Instructions*

All machinery must be accompanied by instructions in the official Community language or languages of the Member State in which it is placed on the market and/or put into service.

1.7.4.3. Sales literature

Sales literature describing the machinery must not contradict the instructions as regards health and safety aspects. Sales literature describing the performance characteristics of machinery must contain the same information on emissions as is contained in the instructions.

Today's manufacturing environment is becoming more and more customer-focused. The most recent manufacturing process innovation is called "agile manufacturing," in which the focus is on responding quickly to customer needs and a changing marketplace. In some industries this means that many products are designed to be customized, with the buyer choosing among various options and no one "standard" model. Naturally, the manual needs to be just as agile: if the machine is customized, the manual must be as well.

"HEY, ALEXA!" TECHNOLOGY PERMEATES EVERYTHING

Certainly, we already live in a technological age. Newspapers are disappearing because most people get their news online. Music is streamed on demand, making CDs seem almost (but not quite!) as antiquated as cassette tapes. Only professional photographers and serious amateurs buy cameras anymore, because the digital cameras built into smartphones can produce high-quality photos sufficient for most of our needs, from documenting a family vacation to capturing meeting notes from a whiteboard or scanning a receipt. Our cars have built-in sensors that tell us when a tire needs air or when we are too close to the car in front. And the GPS built into vehicles and smartphones can give us turn-by-turn directions to almost anywhere, making unnecessary the struggle to re-fold a road map.

ALTERNATIVES TO REALITY

Two developments have already changed the landscape for manufacturers: virtual reality and augmented reality. The first refers to technology that allows a user equipped with a special headset and gloves to be immersed in a fully three-dimensional artificial environment with which the user can interact in a seemingly natural and "real" way. Some manufacturers of industrial equipment are already using virtual reality to train machine operators. It's just like working with the actual machine—but safer. The second is augmented reality, which also requires equipment—usually in the form of a smartphone or special glasses. With augmented reality, the user is interacting with the actual environment, but has automatic instant access to additional information. For example, a maintenance engineer at an industrial plant could have an assembly diagram or a complete service record overlaid on the actual piece of machinery being serviced.

THE INTERNET OF THINGS

But wait—there's more! Already beginning to revolutionize manufacturing and human interaction with products is the coming "Internet of Things," a term for the concept of a wide variety of objects—from machines to appliances to clothing to roads to human organs—being connected to the Internet, and therefore, potentially connected to each other. Driven by a number of advances, including cloud computing, which makes data instantaneously accessible to multiple users in different locations; small, inexpensive sensors that can monitor performance and environmental factors; big data and analytics; and Radio Frequency Identification (RFID) tags that can be attached to almost anything, the Internet of Things will allow just about anything to be connected to just about anything else—with or without human intervention.

On the consumer level, perhaps your refrigerator will sense when the milk is running low and add milk to a shopping list that appears on your smartphone. Some aspects are already here: with the right apps and "smart" devices, you can turn up your home thermostat from your desk at work or send a temporary code to a repair technician to unlock your front door. If you have a voice-activated virtual assistant like Amazon's Alexa, you can already program it to control a sequence of smart-home functions on a single command, so that telling the device "good night" will cause it to lower the thermostat, lock the doors, and turn off lights. Wearable or embedded biometric devices will automatically monitor various health functions and provide real-time information to your physician or automatically adjust the dosage for an insulin pump. On an industrial level, machines will be able to adjust performance automatically to varying conditions, self-diagnose malfunctions, and even direct a 3D printer to produce replacement parts and a robot to install them—all without human direction.

Naturally, such a radical transformation will bring with it opportunities and challenges, including for writers of manuals and warnings. This transformation, sometimes called the "4th Industrial Revolution," promises safer products, perhaps even products that will not *allow* a user to operate it in an unsafe way. Maybe your table saw will check to see if you are wearing your safety glasses (either with a built-in camera using facial recognition software or simply by communicating with the RFID tag in your safety glasses using near-field communication) and refuse to start unless you put them on. Will there still be a need for a warning to wear safety glasses?

On the other hand, the potential also exists for injuries and damage caused either by imperfect machine-to-machine (M2M) communication or by intentional hacking or sabotage of connected devices. Currently we tend to think of cyber security in terms of things like invasion of privacy or fraud by identify theft. But what if a hacker could turn off your furnace in January and cause your pipes to freeze and burst? Or disrupt the sensor-based data system that forecasts when a bridge or airplane wing is about to fail? Or scramble the real-time tracking system for parts and shipments and disrupt a supply chain?

A product manual typically is intended to tell a human how to operate or maintain a stand-alone product. How will that change when the product is capable of being connected to a wide variety of other products that can affect its operation? What safety concerns will need to be addressed and how can that be accomplished when the writer of the manual cannot predict what product connections will be made?

Without doubt the coming Internet of Things will bring much change and probably considerable disruption. The job of technical writers producing product documentation will change as well. And yet . . . until such time as products become completely intuitive or totally autonomous, we will still need to provide users with instructions for operating and maintaining them. The form those instructions take may change, but the basic principles of good communication will not.

THE GLOBAL VILLAGE

Thirty-five years ago, relatively few U.S. companies sold products overseas; now the majority do. For many companies, products are not only sold to foreign markets, they are produced internationally as well. This change brings with it many opportunities for expanded markets and flexibility in production, but it also brings challenges. Not the least of these challenges is producing product documentation.

In the last decade, hundreds of companies have established factories outside the United States to benefit from lower tariffs, lower labor costs, and other business advantages. These multinationals fall all along the spectrum from global giants like Nike to small mom-and-pop companies producing specialty products. Regardless of size, these companies represent a significant piece of the economic pie. As of 2016, U.S. multinational parent companies employed more than 22% of all U.S. residents employed in private industry and produced value added of $3.9 trillion—or nearly a quarter of total U.S. private industry value added.[2] Any time a company has factories in different locations, ensuring consistency and quality is a challenge. When the different locations happen to be in different countries, the problem is only compounded.

How do companies deal with this problem? Certainly, the first step is to develop comprehensive and detailed specifications for the product. The next is to establish inspection and quality-monitoring systems, and the final step is to ensure that a feedback loop exists to fix any problems found. Many companies have a process like this in place. But is the same process in place for documentation? As technological advances make multinational manufacturing the norm, manuals, too, will need to be consistent from one manufacturing location to another.

Commonly, companies produce different product lines in different locations. For example, a vehicle manufacturer may make pickups and SUVs at one location, sedans in another, and commercial trucks in a third. It makes sense for the truck manuals to be written and produced at the truck factory location—after all, that's where the subject-matter experts are—and the same goes for the other locations. How can you ensure that the manuals and warnings (when they need to be) are the same from one to the next? It's difficult enough if the factories are in different states—it's much more so if one is in India, another in France, and a third in Mexico.

In these situations, it is helpful if the technical publications (tech pubs) function is located organizationally in an area that has a company-wide mission rather than a location-specific one. In the case of the vehicle manufacturer, for example, the marketing division might be a better location for tech pubs than say, the engineering division, since the engineering function is specific to each product line, while marketing probably is charged with developing sales and promotion strategies for all product lines. Wherever the tech pubs function is placed organizationally, it must have authority to set company-wide standards for manuals and warnings.

Expanded User Groups

Selling your products overseas means that you must design product documentation that reaches an expanded user group and provides them the information they need to use your product in what may be a very different use environment from what one would expect in the United States. Differences in cultural expectations and traditions and variations in infrastructure can dictate changes to both the content and presentation of information in a manual. Chapter 2 addresses several of these.

If your product is being marketed in a country whose culture is different from yours, be sure to investigate the cultural landscape, but don't accidentally become the insular and insulting "ugly American." Do not confuse unfamiliarity with inability: never underestimate the speed and skill with which people in other countries learn new information and new technologies. Africa, for example, is the world's fastest-growing market for cell phones, growing from about 7.5 million subscribers in 1999 to 420 million in 2017.[3]

BRIDGING COMMUNICATION CHALLENGES

The obvious communication challenge when selling products internationally is differing languages. Even the same language can have considerable variation from one place to another. The French spoken in France, for example, is quite different from that spoken in Quebec. Sometimes these differences occur within a single nation: in China, for instance, people from different parts of the country may not be able to understand each other readily, even though both are speaking Chinese.

If you have a language barrier to overcome, you have two fundamental choices: you can translate your manuals into the language of your users, or you can make the English simple enough that your users can translate it themselves. The first is more common—and indeed, is mandatory for sales to any of the nations in the European Union. The other strategy is used more often when the documents being produced are used by a limited audience. Service manuals, for example, especially for industrial or specialized products, are ordinarily used by factory-trained technicians. Some companies have decided to train their technicians to understand a controlled-vocabulary English rather than translate the manual.

Controlled-vocabulary English strictly limits the words that writers may use to produce manuals. It requires consistent use of technical terms to refer to parts and even requires that ordinary words have only one meaning. For example, the word "close" can mean "shut," and it can also mean "near." The word "check" can mean "test" or "examine" and it can also mean "stop" (as in "check-valve"). In a controlled-vocabulary manual, only one of those meanings would be allowable. In addition, these systems limit the use of complicated grammatical structures that may be difficult for a non-native speaker to follow. Companies that have used these systems include Caterpillar, Ford, IBM, Kodak, and NCR. The Aerospace and Defence Industries Association of Europe (ASD) has adopted a particular version of controlled-vocabulary English called Simplified Technical English and defined it in an association specification: ASD-STE 100.

Translation problems are a staple of discussion at technical writing seminars. Everyone has their favorite examples of instructions that have been turned into English gibberish. Here's a particularly egregious example from the box of a food processor manufactured in China:

> In order that the article has minced could be perfectly cut. Knocked Vigorously on the bud Superior hand Opened.

What in the world does that mean? Whatever was intended, the result is useless, and readers are left to figure out on their own how to use the product. Nor is it a one-way issue: no doubt at this very moment, someone in China is chuckling over an equally incomprehensible instruction that has been mistranslated from the original English. As technical writers, what can you do to ensure good translations and usable formats? This chapter offers a variety of strategies to accomplish those goals, recognizing that, like much else in manual design, trade-offs are inevitable. But first, when do you need to have a manual translated?

The first edition of this book did not consider that manuals for products sold solely in the United States should be translated, although it did say that "bilingual labeling is desirable . . . for a number of products, especially those used in agriculture . . . where the labor force may be Spanish-speaking." These days, most products sold in the United States are accompanied by instructions in (at least) English and Spanish. The fastest-growing ethnic segment of our population is Hispanic. According to the U.S. Census Bureau, the Hispanic population in the country as of this writing is almost 57 million people—and by 2050 that number is predicted to double.[4] Now, not all of those people speak only Spanish—but a significant number do. As of 2016, more than one in ten U.S. residents (13.3% or approximately 40 million people) spoke Spanish at home, although more than half reported that they also spoke English "very well."[5] That leaves something less than one-half who do not speak English very well. Nor are Spanish-speakers concentrated in agricultural jobs anymore. Many industries, from construction to hospitality to food processing and more employ large numbers of Hispanic workers.

Most immigrants learn English relatively quickly—but they cannot become instantly fluent. Manufacturers have an obligation to provide proper instructions and warnings to enable their customers to use their products safely, and if a significant percentage of your customers are likely to be Spanish-speaking, it just makes sense to provide a Spanish-language manual as well as an English one. Are you legally obligated to provide a Spanish manual for sales in the United States? No—but you might have to defend a decision not to, particularly if you have marketed your products to Hispanics with Spanish-language ads or brochures.

If you sell your products in Canada, you must provide instructions and warnings in French as well as English—that is a requirement under Canadian law. If you wish to sell your products in the European Union, you will need to obtain the CE marking. To do so normally requires translation of instructions and warnings into one or more of the 24 official languages spoken in the 28 countries that make up the EU Member States. And if you sell to Africa, Asia, or the Middle East, you have a host of other languages to deal with.

You may have to translate more than words. English or SI units will probably need to be converted to their metric equivalents. And even the numbers themselves may have to be "translated." In Europe, numbers are punctuated the opposite from how they are in the United States. For example, the number 2,000,158.025 in the United States becomes 2.000.158,025 in Europe.

Furthermore, the manuals may need to be "localized" as well as translated. Localization refers to the practice of making the manual contents appropriate to the local culture and experience. As a simple example, translating a warning about plugging a power cord only into a properly grounded receptacle can be translated easily enough, but if the accompanying visual shows a North American three-prong plug and the product is going to Ethiopia or Syria, the visual will not make sense, because the equivalent plug there has three round pins set in a straight line rather than the North American style of two vertical flat blades with a round grounding pin centered below them.

Natural language is rich, slippery, and laden with nuance. In the passage from one language to another, the meanings of words are sometimes skewed or miss the mark entirely. For example, an English manual which read, "Secure the 5/8-inch bolt" was translated into German as "Put the 5/8-inch bolt behind bars." The word "secure" was completely misunderstood. Sometimes mistranslation merely produces howlers, like the translation of "hydraulic ram," as "water goat." However, mistranslation becomes serious business when safety and precision are at issue.

We tend to think of translation as transferring the precise information contained in a passage of text as accurately as possible into a different language. Thus, in the previous example, the word "secure" was mistranslated as "put behind bars" rather than "ensure [the bolt] remains in place." But sometimes, rendering accurate meaning is not enough, just as telling the truth may mean more than just not lying. All natural languages operate within a cultural context. In the same way that tone of voice and body language carry meaning in a conversation—meaning that can be very different from culture to culture—how information is expressed in writing can also carry meaning. James Melton Jr. researched the competencies needed for translating in a training context. He draws a distinction between translation and interpretation, in which translation means getting the words right and interpretation means getting the intent right.[6]

As the global marketplace has expanded, so have translation options. Most companies now use commercial translation services. These firms offer multiple languages and reasonably quick turnaround times. They often use native speakers for whom English is a second language, rather than the other way around. These people have an instinctive feel for what works in their own language—but they have studied English, so they also have a conscious understanding of English nuance.

Be prepared to pay for language skills. Some translators charge by the word and others charge by the page or job, depending on the nature of the original. Spending $50,000 or more on a translation is not unheard of. Most of the costs for translated manuals, of course, are in that first copy. You can't avoid the cost entirely, but you can reduce it. Other than finding translators who will work for free (good luck with that approach), these three tactics will help you minimize the cost of translating your manuals:

- Use a modular approach.
- Minimize the text that needs to be translated.
- Make the remaining text easy to translate.

Each of these can be used independently, of course. Putting all three together yields the best return.

Chapter 3 discusses organizing information into more or less stand-alone modules that can be used in different manuals. For example, the section that explains the principle of positive-pressure ventilation can be used in several different manuals for products that warm, dry, humidify, or otherwise condition room air. In the same way that you only have to write the section once and then can use it in different contexts, you only have to translate it once for each language the manuals will need. The more that you can design your manuals to rely on this mix-and-match modular idea (which, incidentally, is ideally suited to agile manufacturing), the less you have to pay for repeated translation of essentially the same content.

Naturally, the less text there is, the less text needs to be translated. Some of the general writing strategies discussed in Chapters 3 and 4 that produce tight, concise prose also help to minimize translation cost and make translation easier. Not only will there be less text overall, but what remains will also probably be clearer and more precise. On the other hand, the more you can rely on visuals to carry the message, the better. Chapter 6 explores the pros and cons of the "no-words" manual. While it's not a perfect solution, a pictorial-only manual certainly keeps text to a minimum. Be aware, however, that even visuals may have a cultural context that needs addressing. For example, as Patrick Hoffman notes, "Never use symbols of the hand or fingers in your documents, like the warning 'hand' or the reminder 'finger' as they may offend audiences of some Mediterranean-bordering countries and cultures."[7] Be especially careful of any pictorials that involve images that carry symbolic value in the culture.[8]

PROLIFERATING STANDARDS AND LAWS

As industrialization spreads across the globe, standards and regulations relating to products and product safety have proliferated. One of the challenges for companies selling outside the United States is ensuring that their products comply with these. The requirements vary from country to country, and sometimes they conflict with each other. Simply learning what the standards are and keeping up with periodic changes can be a full-time job for more than one person. A complete explanation of the global regulatory and standards environment is far beyond the scope of this book, but a look at some of the most important ones that are likely to affect many manufacturers is appropriate. Since the most common overseas export target for U.S. products is Europe, we begin with the European Union.

EUROPEAN UNION REQUIREMENTS

The European Union (EU) began with the signing of the Treaty of Rome in 1957. The six signatories (Belgium, France, the Federal Republic of Germany, Italy, Luxembourg, and the Netherlands) sought to "lay the foundations of an ever closer union among the peoples of Europe."[9] Since then, the EU has grown to 28 Member States, and the governing bodies of the EU have enacted numerous directives, annexes, and other legislation. For those products covered by EU directives, obtaining the CE marking that indicates compliance with EU requirements is a precondition for selling in EU countries.

Most of the regulations have to do with the technical aspects of products, but some specifically address instructions and warnings. Probably the most widely known of these is the Machinery Directive, originally published in 1995, amended in 2006, and further amended in 2009 and 2013.[10] This very broad directive covers a wide variety of products, ranging from circular saws to locomotives. Much of the directive deals with physical aspects of products (e.g., characteristics of safety equipment), but Section 1.7 of Annex 1 specifically addresses the instructions and warnings accompanying the product. Section 1.7.4 states that: "All machinery must be accompanied by instructions in the official Community language or languages of the Member State in which it is placed on the market and/or put into service."[11] Subsequent sections of the Annex specify the content to be included in the manual.

One of the resolutions of the EU Council is of special interest, in that it sets forth principles for good instructions. It is the "Council Resolution of 17 December 1998 on operating instructions for technical consumer goods."[12] This resolution encourages member nations, manufacturers, business associations, and consumer associations "to pursue the objective of making information available to consumers, enabling them to make safe, easy, proper, and complete use of technical goods" and "to consider, for example, the possibility of voluntary agreements between manufacturers and consumer associations on the design and content of operating instructions and product labeling and award schemes designed to foster the introduction of state-of-the-art, consumer-friendly operating instructions."

The resolution then lists what makes "good operating instructions" in seven areas:

- Development of instructions for use (including standards, guidelines, usability testing).
- Content (basic sections of manuals).
- Separate operation instructions for different models of the same product.
- Safety instructions and cautions.
- Language of manuals.
- Communication of information (clear, precise, user-friendly, particularly for groups such as the elderly).
- Storage of operating instructions for future reference.

You can see how user-friendly the EU documents themselves try to be, and how the standard the EU is trying to encourage for manuals is simply good writing/design advice.

One of the clear messages from the various EU documents is that the "product" is not the manufactured item alone, but rather the manufactured item together with all relevant instructions, warnings, and even sales literature. The EU articulates that the manual is not to be treated as an afterthought, but rather as part and parcel of the product itself.

Another clear message from the EU is that multilingualism is encouraged—there seems to be no movement to reduce the number of official languages. Instead, the push in the "New Approach" is harmonization of standards and encouraging use of accepted symbols to convey crucial information. As an example, Section 1.7.1 of Annex 1 of the Machinery Directive states, "Information and warnings on the machinery should preferably be provided in the form of readily understandable symbols and pictograms." Challenges related to translation are not likely to decrease in the foreseeable future.

INTERNATIONAL STANDARDS

Unlike regulations, which have the force of law, standards encourage voluntary compliance. Organizations like the American National Standards Institute (ANSI), the International Organization for Standardization (ISO), and the International Electrotechnical Commission (IEC) publish hundreds of standards relating to various aspects of all kinds of products, from motorcycle helmet crashworthiness to the design of lift systems. Usually, these standards are developed by a committee of individuals representing a variety of public and private interests in a given industry or product type. These are consensus standards, meaning that they describe what are generally considered best practices. While compliance is in theory voluntary, in practice it may be mandatory, because some governmental regulations require adherence to one or more standards, incorporating them by reference.

Two international documents address writing instructions: IEC 82079–1 and ISO/IEC Guide 37:2012. Both provide solid advice for preparing usable product manuals and instructions. IEC 82079–1 *Preparation of instructions for use* nicely reflects the expanded definition of "product" discussed earlier: Section 4.1.2 is titled "Instructions for use are part of the product." Both standards emphasize the importance of making sure that all product literature is consistent and following the principles of clear communication. IEC 82079 covers manuals and instructions for "all types of instructions for use that will be necessary or helpful for users of products of all kinds, ranging from a tin of paint to large or highly complex products, such as large industrial machinery, turnkey based plants, or buildings."[13] In 2012, ISO published ISO/IEC Guide 37:2012 *Instructions for use of products by consumers*. This guide lays out sound principles for designing and writing effective instructions, paying special attention to communicating safety-related information. Among other things, it recognizes the importance of designing instructions to work for all types of users, noting specifically these characteristics:

- Age
- Gender
- Cultural background
- Capabilities (novice vs. skilled, those with disabilities or low literacy levels)

The Guide also calls for instructions to be evaluated by an independent expert, a panel of users, or a combination of both, and it provides in Annex A checklists for content and communication effectiveness.

Two other standards technical writers should know about are the ISO 9000[14] and the ISO 14000 series. The first is a family of standards defining a quality management system and the second is a family of standards defining an environmental management system. The first is relevant to technical writers because quality standards are often confused with product safety standards. The second is relevant because it requires certain labeling that technical writers may become involved in developing.

Quality is not the same as safety. Quality refers to whether the product meets an acceptable level of consistency and whether it meets its customers' expectations. A chain saw with no chain brake and a round-ended bar (no kickback reduction) could be very high quality—having a low rate of parts out of specifications and performing well according to what the customers want it to do, such as cutting efficiently, starting reliably, etc. But it would not necessarily be reasonably safe. Conversely, a product could be quite safe—few or no associated hazards, adequate warnings and instructions—but also of low quality. When a company gains ISO 9001 certification, it demonstrates that it has put in place systems to ensure management of quality and continuous improvement, but it says nothing about safety.

Similarly, ISO 14000 is concerned with companies' environmental management systems, not directly with the general safety (or environmental safety) of the manufacturer's products. This family of standards deals with aspects such as documentation and setting and achieving environmental goals, among others. But just as the Incident Command System used in emergency services sets forth an organizational framework for responding to an incident, but doesn't tell you how many gallons of water to pump on a fire, ISO 14000 provides a framework for managing environmental issues, but does not set maximum emissions standards.

Both of these are international standards, which is to say that they are widely used throughout the world—although less so in the United States than elsewhere. However, both are gaining ground in the United States.

Law and Regulation

The European Union directives take the place of country-specific requirements for the Member States of the EU. This is very good for manufacturers, because once their products are shown to conform to applicable EU regulations, they can be freely sold throughout the 28 countries that are part of the EU. That only leaves 165 nations to go.[15] Fortunately for manufacturers, most of the rest have less-stringent requirements (or in some cases, no requirements at all). In general, any requirements that various countries set out for products have to do with the manufactured items themselves, and not the manuals that go with them.

Some countries maintain websites that explain trade requirements. An excellent example is Japan's JETRO (Japan External Trade Organization) site (www.jetro. go.jp/en).

The standards that affect international trade continue to develop. Selling your products overseas requires continual research to keep up with the shifting landscape of international standards and regulations. Most companies, unless they are large enough to have an in-house expert, use consultants to guide them in how to comply with international requirements.

In addition to meeting standards and complying with regulations, manufacturers must be prepared for evolving products liability law. As we shall see in Chapter 8, products liability law is complex and variable within the United States, let alone in the rest of the world. The good news is that if your products meet regulatory requirements (such as EU mandatory standards) and would be considered "reasonably safe" for sale in the United States, they are probably acceptable internationally as well. Products liability law is more developed—and more litigation takes place—in the United States than anywhere else. As one the speakers at a recent products liability conference put it, "If you're going to get sued, it will be in the U.S."

Nevertheless, several countries, including Japan, China, and Taiwan have put in place products liability statutes. Most of these international laws, including the EU's Product Liability Directive, adopt the principle of *strict liability*, meaning that the manufacturer (or other seller) is liable for injuries caused by product defects, regardless of whether the manufacturer was negligent. The focus is on the product, not the conduct of the manufacturer. Some laws impose a time limit for bringing actions or provide a defense if the manufacturer was unaware of the defect at time of sale. We can expect that products liability litigation worldwide will increase over time.

SUMMARY

The French have a phrase for it: *plus ça change, plus c'est la même chose*. It means, roughly, "the more things change, the more they remain the same." The world of technical writing and product documentation is evolving at a dizzying pace. Yet until products become autonomous, they will still need to be accompanied by instructions. We still have manuals because they still serve important purposes for the customer and the company. For the customer, they explain how to use the product safely, serve as an ongoing reference, and as a safety resource. For the company, they function as needed documentation, can be a product development resource, and serve as the face of the company to the user.

But not everything remains the same! Customers have much higher expectations for products and manuals than they did just a few years ago. Driven in part by products liability litigation, customers expect products to be safer and documentation to be accessible and easy to use. The manual is now considered to be an integral part of the product, along with sales and marketing materials. Technology has changed how we get information (and how instantly we expect it) and is beginning to change how we learn about and interact with products. Augmented reality has made it possible to access product-related information automatically in real time. Virtual reality has made it possible for operators to learn to use complex industrial machinery without fear of injury or damage. The Internet of Things will change the landscape once again, altering how humans interact (or don't) with products and how products interact with each other. These changes mean that the content and form of instructions are changing too. A paper manual packed in the box with the product may be replaced by a QR code that takes you to a .pdf download instead.

Whatever form the documentation takes, one of the biggest changes is that it now must serve a global audience. For manual writers, that global expansion has brought with it many challenges, including cultural differences, the need to translate manuals (or find ways to develop instructions that don't need translation), and the need to take into account wide variations in conditions of use and availability of infrastructure support. Product safety is a worldwide concern, and new standards and regulatory requirements have emerged to address it, including standards for product documentation. At the same time, with the explosion of global commerce, the spectrum of users for a given product has dramatically increased. Knowing who is using your product as well as knowing its potential hazards are key to developing documentation that will serve to keep users safe and reduce liability. Those key tasks are the subject of the next chapter.

NOTES

1. Bureau of Labor Statistics, www.ncbi.nlm.nih.gov/pmc/articles/PMC3151179/#B1, accessed October 01, 2018.
2. BEA 18–41 news release, August 24, 2018, Bureau of Economic Analysis, U.S. Department of Commerce.
3. Sharon LaFraniere, "Cellphones Catapult Rural Africa to 21st Century," *New York Times*, August 25, 2005. Also www.africanews.com/2017/07/25/over-half-a-billion-mobile-subscribers-in-africa-by-2020-hi-tech//, accessed October 13, 2018.
4. United States Population Projections: 2000–2050. United States Census Bureau, 2009, Revised March 19, 2018, www.census.gov/library/working-papers/2009/demo/us-pop-proj-2000-2050.html, accessed October 16, 2018.
5. United States Census Bureau, Newsroom Facts for Features 2017, www.census.gov/newsroom/facts-for-features/2017/hispanic-heritage.html, accessed October 16, 2018.
6. James H. Melton Jr., "Lost in Translation: Professional Communication Competencies in Global Training Contexts," *IEEE Transactions on Professional Communication* 51, no. 2 (2008): 209.
7. Patrick Hofmann, "Localising and Internationalising Graphics and Visual Information," *IEEE Transactions on Professional Communication* 50, no. 2 (2007): 91–92.
8. Guang Han and Jingxin Lin, "Cultural Symbols or Taboos: The Cultural Conflicts Reflected in the Cultural Image in International Advertising," *China Media Research* 2, no. 2 (2006): 38–43.
9. Preamble, Treaty of Rome.
10. The Machinery Directive 2006/42/EC was published on 9 June 2006 and became applicable on 29 December 2009, and was further amended in 2013. Europa, https://ec.europa.eu/growth/sectors/mechanical-engineering/machinery_en, accessed October 17, 2018.
11. While this seems to indicate that translation would not be required, since English is an official language of the EU, in actuality translation is often required by the individual Member State's treaty governing its admission to the EU.
12. 98/C 411/01.
13. IEC 82079–1 Section 1: Scope, p. 8. International Electrotechnical Commission Edition 1.0 2012–08. ISBN 978–2–83220–096–4.
14. ISO 9000 is a series of standards. Most are guidance documents, but one, ISO 9001:2025 sets out the requirements for an organization to become ISO 9001-certified.
15. Based on the current membership of the United Nations.

2 Know Your User and Your Product's Hazards

OVERVIEW

Chapter 1 laid out several functions that manuals perform for the customer and the company. All of these functions serve the overarching goal of enabling people to use and (sometimes) maintain the company's products properly and safely. As products liability litigation focuses more on instructions and warnings, manuals become more important in helping to improve safety and reduce liability. Among the first steps to creating effective manuals are identifying the users of the product and its manuals and understanding the hazards associated with the product. Both must be addressed early on in manual development to ensure that the instructions and warnings accomplish the goal of keeping the user safe.

KEEP THE USER SAFE!

Manufacturers have a duty to provide a "reasonably safe" product and to provide adequate instructions for safe use. How safe is reasonably safe? Unfortunately, it is impossible to provide a clear, black-and-white answer to this question. What is considered reasonably safe depends on several factors and often evolves over time as expectations change and lawsuits are decided. In general, a reasonably safe product provides an *optimal* level of safety. Optimal safety is not absolute safety. As the *Restatement of the Law Third, Torts: Products Liability* puts it, "Society does not benefit from products that are excessively safe—for example, automobiles designed with maximum speeds of 20 miles per hour—any more than it benefits from products that are too risky."[1]

Is My Product Reasonably Safe?

Courts often consider these factors when evaluating whether a product is reasonably safe:*

- Risk vs. utility
- Available alternatives
- Consumer expectations

---

* Note: I am not an attorney and I do not give legal advice. The references to legal requirements in this book are broad generalizations about a very complex area of law that includes considerable state-to-state variation. They are intended to be informative only.

Risk vs. Utility

It is hard to imagine a product without some potential to harm. Even something as innocuous and commonplace as an ordinary pencil can be lethal if used as an unconventional weapon to stab someone. Yet that small risk is perfectly acceptable, given the general usefulness of pencils as writing instruments. Because most products involve potential hazards, part of a good product development and improvement practice is to identify hazards that are present in a proposed or existing product and try to find ways to reduce those hazards *while retaining the usefulness of the product*. We will look in depth at how to do that later in this chapter.

That a product is dangerous does not mean that it is *unreasonably* dangerous. Some inherently dangerous products—chain saws, for example—are also very useful. Chain saws have unguarded cutting teeth; they sometimes "kick back," potentially causing the user to lose control of the saw; and gasoline-powered models emit pollution. However, chain saws are without question useful tools and sometimes may be the only tool suitable to accomplish a particular task. The fact that a chainsaw is inherently dangerous—in other words, it poses a significant risk of injury to the user and/or to bystanders—does not in itself mean that the saw is unreasonably dangerous. The utility of the saw makes the assumption of some degree of risk reasonable.

Available Alternatives

A product might be considered unreasonably dangerous in some contexts but not in others. A propane torch would be an unreasonably dangerous tool with which to light a candle, but (in the hands of a skilled plumber) a reasonably safe tool to use to sweat-solder copper pipe. The difference is simply that there are safer alternatives readily available for lighting candles (matches, cigarette lighters), but not for soldering pipe.

Consumer Expectations

A third factor to weigh in determining whether a product is reasonably safe is consumer expectations. As noted in Chapter 1, as products have become safer over time, people's expectations for safety have increased. Years ago, chainsaws were sold without chain brakes, lawn mowers without dead-man switches to stop the blade if the operator's hands left the handle, and tractors without rollover protection. A 1955 Chevrolet Bel-Air was considered state-of-the-art at the time, despite having no passenger restraint system, an unpadded metal dashboard, and neither antilock brakes nor traction control, let alone a back-up camera.

What Is "Reasonably Safe" Depends on Who's Using the Product

The level of safety required for a product to be considered reasonably safe varies not only over time, but also with the anticipated user group. Whether a product is reasonably safe may depend on who's using it. An excimer laser used to reshape the cornea of the eye in LASIK surgery might be unreasonably dangerous in the hands of an untrained consumer, but reasonably safe when operated by a highly trained ophthalmologist.

To design and write effective instructions and warnings, the writer must know who the users are and know what the product's hazards are. Both are equally important.

Remember that under strict liability, poor instructions and warnings can be considered a product defect, just like an unsafe design—and can expose the company to liability. A word of caution is in order: writing to prevent liability can produce manuals and warnings that read like the fine print in an insurance policy or disclaimers that put the burden of preventing injury onto the user. A much better approach is to focus your efforts on writing instructions and warnings to *protect the user from injury or harm*. After all, if no one gets hurt and nothing bad happens, no one will sue!

What's the difference between writing to prevent liability and writing to protect the user? Consider the following examples, both warning the operator of an x-ray machine about radiation exposure.

Example 2.1: Warning Designed to Prevent Liability

All persons authorized to use this equipment must be cognizant of the danger of excessive exposure to x-radiation, and the equipment is sold with the understanding that the ABC Company, its agents, and representatives have no responsibility for injury or damage which may result from exposure to x-radiation.

Example 2.2: Warning Designed to Protect the User

WARNING! This equipment emits x-rays. Excessive exposure to x-rays can cause illness, including cancer. Limit your exposure by wearing appropriate PPE and following safety procedures. Wear a radiation monitoring badge when using this equipment. See manual for more information.

Example 2.1 does not give the user much useful information. It indicates that excessive radiation exposure is dangerous, but does not explain why. It provides no specific actions to take to avoid excessive exposure. In essence it is saying, "This equipment does something that's hazardous, and it's your responsibility to understand the hazard and what to do about it. If anything bad happens, it's not our fault." Example 2.2, on the other hand, provides specific hazard consequence information (illness, cancer) and offers three specific actions to take to prevent over exposure. Finally, it points the user to the manual for more information.

Under strict liability, the disclaimer approach in example 2.1 might not provide much protection for the manufacturer anyway, but if there were a lawsuit, Example 2.2 would certainly make the company look much better to the jury. Example 2.2 makes it clear that the company is making a good faith effort to provide the information needed for the operator to stay safe.

WHO ARE THE USERS (AND WILL THEY READ WHAT I WRITE?)

Knowing who is using your product will help you write a manual that serves their needs—providing they actually read the manual. Given that people generally don't like to read manuals, you might reasonably ask, "Is anybody going to read what I'm sweating bullets to write?" The answer, happily, is yes. People do read manuals. In fact, according to a study by Karen Schriver, documented in her landmark book *Dynamics in Document Design*, only 4% of those studied *never* read manuals.

"Seventy-nine percent of users report they would buy from a company they thought had clear communications," and "more than half of participants would consider paying more for a product if they knew it had a clear manual."[2]

People do read and use manuals—they just don't use them as we expect or, frankly, as we wish they would. Some of what we find is the following:

- People almost always would rather ask a person than read a text. People read only if they can't get information in another, easier way. And when they do read, they read as little as possible.
- People will use manuals only if the reward meets or exceeds the effort.
- People read with memory—their past encounters with instructions shape their present experience.
- People read a manual while they are doing something else—typically, while working with the product. This simultaneous activity is crucial to understanding how to design a manual to help a user.
- Many people are reading to use a product on the job. They are not reading for fun; they are reading only to use a product to get something else done.
- People are often in a hurry, under stress, anxious, or at least uncertain when they pick up a manual, particularly if the product is unfamiliar to them or appears to be technically sophisticated—or if it involves software of any kind.[3]

Our job is to meet that 96% who do read manuals where they *are*—not where we wish they were. For all the reasons discussed in Chapter 1, customers need manuals—our job is to find out who our readers are and what it is that they need and want. Then we can design a manual they will use—because the manual works to help them use the product, and the reward exceeds the effort.

WHY DO I NEED TO FIND OUT ABOUT MY AUDIENCE?

Manual writing is fundamentally a design problem, and just as with designing a desk or a machine or a piece of software, decisions are driven by two questions:

- Who will use it?
- What will they use it for?

For example, it seems so obvious that when we set out to design a school desk, we need to ask whether it will be used by a first-grader or a college senior. Is the calculator app we are designing intended for an engineer, an accountant, or a carpenter? For tangible products and software applications, user characteristics necessarily guide the design. The same is true for manual users. Using a manual or any other instructional text is a complex task. It involves at least four elements:[4]

- The instructions themselves
- The product
- The user's knowledge of the product
- The user's abilities and aptitudes

If the writer of the instructions misjudges the user's prior knowledge or abilities, the instructions will not be as effective as they might be—and the user may think that reading them is more trouble than it's worth. Unlike the college freshman trapped in a boring lecture, product users won't stick it out (or sleep through it)—they'll find another way. Too often, that other way can lead to product misuse and potential injury.

You need to know everything you can find out about your readers that will affect how you design the manual, because information about your audience should guide a host of decisions:

- What language level should I use?
- Can I assume familiarity with industry technical terms?
- Will my readers prefer more text or more illustrations?
- What languages do I need to present the manual in?
- How much background information should I include?
- What size font should I use? What typeface?
- What kind of paper should I use—or should the manual be solely online?
- Should I create one manual or several targeted to different audiences?
- What size should the manual be?

So, how can you begin to identify your potential readers and learn about them so that you can design a manual that will be worth their time?

SOURCES OF USER INFORMATION

How can you find out who your users are? The answer to this question varies with the product you manufacture. If your company builds million-dollar customized paper converters for the papermaking industry, chances are you have records of exactly who has purchased each one—although the purchasing agent is probably not the user. Even so, you have a good place to start research on who will actually be using the manual. On the other hand, if your company manufactures food processors or electric staplers or pressure washers, the problem becomes more difficult, because your user could be anyone, anywhere. Let's start with the easier problem. Some of the sources you can tap for information about your industrial users include these:

- Sales records
- Sales and service personnel
- Marketing reports
- Product development reports
- Trade journals

Sales Records

If your company makes large capital equipment, begin by checking sales records. You may find that you sell to only a few markets, which will make your job that much easier. You may be able to contact customers directly and conduct formal or informal surveys with users to help guide your manual design. On the other hand, you may find that you have a broader market segment that will be more difficult to assess. Repeat customers can be a source of information about what you are doing right. One-time buyers might be less than completely satisfied—or they simply might not need a second product. Sales records can lead to direct contacts; but even without direct contact, they can help paint a picture of who your manual users might be. Are they primarily in large companies or small? Are they geographically clustered or dispersed? Are they in one industry or many?

Sales and Service Personnel

The sales and service departments can offer the technical writer a wealth of information—and they are all too often resources left untapped. Unlike most technical writers, who spend their days staring at a computer screen in a cubicle, sales and service representatives go out and visit customers. They see the company's products being used in the context of their normal environment. They hear customer feedback and they know what goes wrong when instructions aren't perceived as being clear. They know the shortcuts that users take and they often get an earful when the customers aren't happy with the manual. Generally, however, sales and service reps do not forward this information to the technical publications department. It's not their job. Their job is to generate leads, sell products, give advice on proper use, repair malfunctioning equipment, and sometimes identify when a technical bulletin needs to be issued. Rarely will any of this information filter back to tech pubs.

What's the solution? Simple: put your computer on standby, emerge from your cubicle, and go talk to the sales and service reps. Buy one of them a cup of coffee at the company cafeteria and let him bend your ear for a while. That simple gesture can reap enormous benefits for you and the company. You can cultivate a new friendship that may result in that sales or service rep picking up the phone or shooting an email that lets you know about a problem spot in product documentation. That cup of coffee could potentially even save a life.

Some years ago, I consulted in a products liability lawsuit involving a large grain-storage bin. The bin had been installed on a metal framework so that trucks could drive right underneath it for easy loading. One day, the framework collapsed, causing the full grain bin to fall and crush the truck (and driver) underneath. The bin manufacturer was sued. One of the allegations was that the manufacturer failed to warn of the dangers of elevated installations or to provide specifications for proper supporting structures. The manufacturer's initial position was that such an installation was a misuse of the product and was not foreseeable (more about these concepts in Chapter 8), and therefore the manufacturer had no duty to warn. In essence, the manufacturer took the position, "If we didn't have any idea that someone would do this improper thing in the first place, how could we be expected to warn against it?"

That certainly sounded reasonable—until the manufacturer's sales rep was deposed. He admitted that he knew that customers were installing these grain bins in elevated configurations. If the sales department and the technical publications department had worked more closely together, someone might have thought to address the issue in the manual.

Marketing Reports

Marketing departments live and die by market research. A good marketing department is forever sending out surveys, convening focus groups, profiling purchasers, and so on. Their job is to know who the customers are—that is, who buys the products. Marketing reports can be very useful in locating your users, but need to be used with some caution. Remember that the buyer is not necessarily the user—especially for capital equipment and high-tech medical and industrial equipment. The decision to purchase a particular model computer-based maintenance scheduler that relies on electronic flowmeters, oil monitoring, and other sophisticated electronic sensors may be the responsibility of an engineer in charge of plant operations, but the person who uses it may be the line operator who has lots of experience with machinery, but no formal education beyond a high-school diploma.

Product Development Reports

Products under development are continually reviewed, evaluated, tested, and analyzed to find ways to improve function and marketability. Sometimes the analysis is designed to identify potential problems. Methodologies designed to identify potential failure points within the product, for example, can help engineers predict the consequences of a failure, and quantify the likelihood and risk of harm if a failure occurs. While these studies may not be directly applicable to writing the manual, they can provide the technical writer with a good deal of information about the expected users and manner of use.

Trade Journals

Often, capital equipment and other non-consumer products are geared toward a specific industry. In other words, if you make veneer dryers, you would sell them to wood products companies, not to foundries or fertilizer plants. That not only makes it easier to locate your potential readers, it also makes it more likely that they will have certain similarities—common vocabulary, similar work environment, and similar applications for the products. One of the best sources for information about your users is the trade journals that serve your industry. Every industry has them. If your company manufactures tires, you'll probably find a copy of *Tire Review* somewhere in the office. If your company makes molded plastic components, there may be a copy of *Injection Molding Magazine* in the mailbox. If your company makes pumps used in mines, someone probably subscribes to *Mining Engineering*. These are just a few examples—find out what's available for your industry and start reading. You will gain valuable information about the industry and the people who work in it.

But what if your company makes coffee pots or microwaves or lawn mowers or cordless drills? These and other consumer products have a much broader user base. How can you possibly find out anything useful about your audience? For consumer products, the marketing department is probably one of your best sources of information. Marketing researchers are constantly studying who makes purchasing decisions and what those decisions are based on—and increasingly, web analytics makes it easy to identify who is even *thinking* about buying a product. A quick review of the literature produces a wealth of articles on consumer buying decisions—even cross-cultural studies;[5] but again, marketing focuses on who buys the product, not necessarily who uses the product.

Other Sources

Some of the other sources of information that work for technical writers in industrial products manufacturing will also work for those in consumer products. Certainly, sales information can be helpful. While you probably will not get a list of specific customers, you can learn where most people make their purchases—is it from a focused dealership (such as a farm implement dealer or a hardware store) or from a "big box store" (such as Walmart or Home Depot) selling a wide array of consumer products? Service records, especially in regard to frequent problems or calls for technical assistance, can help identify communication needs. Product development reports can be every bit as useful for consumer products as for more narrow-use products.

WHO ARE THESE PEOPLE?

At the same time that the variety and complexity of products is exploding, the pool of potential users is expanding as well. Knowing pertinent information about your users is essential to writing manuals that meet their needs. Some of the dimensions that technical writers need to consider when trying to define their audiences include these:

- Age
- Gender
- Culture and geography
- Literacy and language

Any or all of these can have important implications for manual design.

Age

The world's population is growing rapidly, but for the first time in history, the average age is also rising. As a result of improved health and declining birth rates, the typical age pyramid, with the largest percentages of the population in the youngest segments, is changing shape. In the United States, people over 85 years of age constitute the fastest-growing age segment of the population. By 2030, about 20% of the U.S. population will be over 65. Even in less-developed nations the trend is similar—although not as far along. In 2007, the over-65 group made up only about 6% of the population worldwide; but by 2050, it is projected to be 15% of the total.[6] An aging population has many implications for manual writers. Beyond the obvious problems that older eyes have with smaller type, there may be more subtle differences in expectations about content and organization, ways of getting and processing information, and willingness to put forth the effort to learn new technology.

The pace of change is furiously fast and ever-increasing. If you think about the amount of change today's 65-year-old has experienced and compare it with the amount of change experienced by a 65-year-old just a couple of hundred years ago, it's simply staggering. When today's 65-year-old was born,

- Television was not yet available in all states, although NBC, ABC, and CBS had begun broadcasting (from 8:00 am to 11:00 pm only) from New York in 1948.
- Photocopiers did not exist.
- Computers, web pages, and email did not exist.
- GPS technology was inconceivable because the only satellite orbiting the earth was the moon.
- All cameras used film.
- Cell phones did not exist and most homes had only one telephone (not yet referred to as a "landline"), that performed only one function—voice calls.

The point is simply that the learning curve for today's adults isn't flattening out—it's getting ever steeper. As the percentage of the population over 65 increases, the need to design manuals and other documentation to meet their needs also increases.

So what do older users need in a manual? The research is limited, unfortunately, and much of it is scattered in the literature of other disciplines, such as psychology or human factors.[7] Some commonalities emerge, however:

- Older readers prefer to have information about the goals and consequences of instructions.
- Older readers may be frustrated or confused by gaps in information that younger readers fill in intuitively.
- Older readers have more difficulty with small font sizes.

Standard advice to the technical writer is to focus on the "need-to-know" information and omit the "nice-to-know." That's good advice—the tricky part is knowing which information goes into which category. The answers may be different for older readers than for younger. Some research suggests that older readers want to know *why* to do an action as well as *what* to do. If the "why" is not provided, they are less likely to perform tasks well.[8] This preference is probably not biologically age-related; more likely it has to do with learned thinking patterns. Regardless of the reason, this "who is your audience" component may have a major influence on what you say in your manual and why you present the information in the first place, especially with respect to warnings.

Prior to the widespread use of computers and smartphones as information portals, information tended to be delivered in fairly linear ways and single dimensions. People read newspapers or they watched a film or they listened to a speech. Each of those modes presented information that was ordinarily given out serially over time. Sure, you could jump from paragraph 3 to paragraph 23 and back to paragraph 15, but there was no logical reason to so, and if you did that, the article would be more confusing, not less. Computers and the Internet have changed all this. Now a chunk of information may be presented as part text, part video, part audible narration— all at once. And hyperlinks provide both a means and logical motivation to jump around. You no longer have to start at the beginning and work your way to the end. You can start anywhere and move anywhere.

Today's 14-year-old knows nothing else; non-linear information acquisition seems perfectly normal, and being forced into linearity probably seems absurdly confining. The same is not true for today's 65-year-old. While he or she may appreciate the value of this kind of flexibility, the non-linear mode is likely to be less comfortable and more difficult to use—it simply doesn't feel natural. The challenge for technical writers is to find modes that work for both groups.

Similarly, older readers may not automatically fill in missing information that is obvious to younger readers. Again, it's not that older readers aren't as intellectually able—it's that their thinking patterns are based on their experience, and in some cases on older technological models.

For example, consider the ubiquitous icons used on an iPhone for various functions. The trash can icon is pretty obvious; slightly less so, but still intelligible is the pencil laid across a square representing a piece of paper (compose a message). But what about the rectangle with an upward-pointing arrow superimposed on it? For an older, novice user, the only experience reference for an up-arrow on a piece of paper is to show North on a map or to indicate "This side up" on a box. It may be patently obvious to a person born in the age of the Internet that tapping that icon will upload the information and send it somewhere (to a printer, to email, etc.), but it's not intuitively apparent in the way the trash can icon is.

As we age, our eyes lose the ability to focus at close range because the muscles that physically adjust the focal point become less flexible. This decline in close visual acuity may be accompanied by other visual problems, such as "floaters" (bits of loose pigment in the eye), cataracts, and more serious conditions such as macular degeneration or glaucoma. Even with reading glasses, type smaller than 10-point may be very difficult to read. The difficulty with small print affects more than just words. Callouts in graphics, the illustrations themselves, icons on screens, all must be large enough for older readers to see easily—or they won't bother trying.

Gender

Forty years ago, if you wrote manuals for power saws, automotive engines, or furnaces, you could be assured that just about all your readers would be male. Similarly, you could assume that the readers of manuals for hair dryers, dishwashers, and electric mixers were almost all female. There were virtually no women operating mining machinery and no men running home sewing machines. That has changed. The breaking down of gender barriers to various occupations coupled with later marriage and high divorce rates means that you must assume that your readers could be of either sex.

What difference does this make? While there do appear to be some differences in the ways that men and women process information (at least statistical differences between large groups of people), the real importance of gender for manual writers is the experience base that men and women draw upon. Similar to the way in which experience differs for older and younger readers, it may vary considerably for male and female readers, although the "experience gap" is narrowing along with the gender gap. Sometimes the different knowledge bases provide fodder for entertaining "battle-of-the-sexes" vignettes, such as the man trying to follow a cooking recipe being stumped by the instruction to "separate the eggs." (Why? If they're together, will they fight?) Or the woman following instructions to install wiring being directed to "be sure the fixture is grounded." (Is it OK to put it down anywhere on the ground?) These amusing images are based on the premise that just from growing up male and female in our society, people have different sets of knowledge. Even though the variation may be diminishing, it's still important to remind ourselves that we shouldn't assume that all of our readers bring the same background knowledge to their reading.

Culture and Geography

As any international traveler knows, navigating cultural differences can be very tricky. Cultural patterns are so ingrained that we often are not conscious of them at all. For example, the distance you stand from someone when conversing varies dramatically from one culture to the next. In the United States and Canada, typical "social" distance is around 24 inches. In other cultures, it may be much less. If someone from another country stands closer than two feet to a North American, it will be perceived as uncomfortably close, possibly even "in your face." To the other person, it's entirely normal. Clearly, the potential looms for misunderstanding resulting from misinterpretation of cultural signals. The same may be true for product documentation.

In the United States (and most Western countries), we assume that the purpose of a manual is to provide information. A secondary purpose noted is that the manual is also a public-relations document that "puts a face on" the company that made the product. In other cultures, these priorities may be reversed. As communication researcher James H. Melton Jr. puts it, " . . . this is a mistaken assumption that Westerners commonly make in countries such as Japan, where relationship building is often the primary purpose of communication and information sharing is secondary."[9] Similarly, we think of our primary audience as the end-user of the product, because in the United States, that's the person who would normally consult the manual (assuming it's available). As another researcher notes, in Mexican factories the instructions are typically read not by end-users, but by trainers or managers who in turn teach the workers.[10]

In the United States, we tend to think analytically—by dividing an item into component parts and understanding how each component functions. We think of an automobile, for instance, as being composed of an engine, drive train, electrical system, cooling system, etc. Accordingly, we would be likely to structure a service manual in just that way, with separate sections for each system. In China, people tend to think more holistically, seeking to understand how the entire assembly functions. Technical manuals in China tend to reflect that holistic view: fewer headings (too divisive), more referrals to earlier material (showing relationship), topic sentences at the end of a paragraph rather than the beginning (inductive vs. deductive order), and so on.[11]

As Mary Lou Fisk notes, "Too often in designing technical documents we fail to recognize that cultural codes inhabit texts, that how we process and accept information is based on world knowledge which we, as representatives of our culture, bring to our work."[12]

This particular area of technical communication research is not yet well developed—each article tends to raise more questions than it answers. So what's a technical writer to do? The same thing you always do: find out as much as you can about your readers and design the manual to fit their needs.

If you are selling in the United States, by and large you can assume a certain shared experience. You may have people with varying degrees of technical sophistication or familiarity with your product, but you can be fairly certain that just about everyone will know what the consistency of peanut butter is, how thick a pencil is, or how big a golf ball is. That shared experience allows you as a writer to make meaningful analogies. You can, for example, instruct your user to stir water into a powder until the mix becomes the consistency of peanut butter. But what if your user has no idea what peanut butter is, much less its consistency?

Similarly, because you can safely assume that any U.S. reader has driven—or at least ridden—in a car, you also can assume that knowing how to use a steering wheel is second nature. If you are writing a manual for a forklift to be sold in Bangladesh or Liberia, however, you might want to include an explanation. Whereas the U.S. has more than 800 motor vehicles for every 1,000 people, in those two countries the rate is only about 3 per 1,000. Even within the western developed nations, experience differences can pop up. A few years ago, I visited Italy with friends, and we stayed in rental apartments rather than hotels, so we could cook some of our meals. In one very nice, modern rental I went to put a casserole in the oven and was confronted with the oven control shown in Figure 2.1. Instead of the familiar verbal settings (Bake, Broil, Clean, etc.), here were unfamiliar symbols. Fortunately, the apartment was equipped with wi-fi, and a quick Internet search saved dinner!

FIGURE 2.1   European oven control using symbols instead of words.

Source: Photograph by author

The conditions of use may be very different from what you might expect in the United States. For one thing, employers in the United States generally have to adhere to strict safety regulations in their workplaces. That is often not the case in the rest of the world. A worker staining furniture in the United States is likely to be in a large, mechanically ventilated shop. In Malaysia or Indonesia, that same operation might take place in a small room with a single open window. For another, the environmental conditions may be quite different. While the continental United States has a considerable range of climate variations, it also has climate-controlled buildings. A machine may work perfectly well in an air-conditioned factory in Phoenix, Arizona, when the outside temperature is 112° F, but it might not fare as well in that same 112° F in Baghdad, Iraq, without climate control. If your product requires special conditions, such as an environment in which temperature and humidity are controlled, are those conditions readily available?

The conditions of use may also include the size of the person using the product. The United States has strict child-labor laws.* Machines that are sized for adults in most cases are used by adults. Other countries may not have such strict laws or, if they do, the laws may not be enforced. Learning that a product intended for use by adults is in fact being used by children may mean that a redesign is needed or procedures modified to accommodate safely a wider range of operator sizes. Even if the product is used by adults, in other countries, the average size of an adult male or female may be smaller or larger than in the United States. For example, on average, men in Japan are about four inches shorter than in the United States, whereas in Denmark, they're about two inches taller.

Another issue for manufacturers is the problem of variations in infrastructure. If your product requires frequent maintenance and parts replacement, does the country to which you are shipping the product have maintenance technicians and parts stores? If repair technicians are not readily available, the user might need to do more maintenance on the product. If parts are not readily available, worn parts may be repaired instead of being replaced. Even if parts can be easily ordered, they may not be so easily delivered if the road system is inadequate. These variations may mean that you need to include more maintenance information in the manual or, if safety would be compromised, warnings against reusing worn-out parts.

Literacy and Language

Literacy levels can also be problematic to predict. According to the United Nations, worldwide about 10% of adults are illiterate. The percentage varies dramatically by region: in East Asia and the Pacific, the overall literacy rate is about 99%, whereas in sub-Saharan Africa, it's only about 74%. Worldwide, men are more likely to be literate than women—about two-thirds of illiterate adults are women.[13] Even in the United States, a 2003 survey[14] by the U.S. Department of Education found that 14% of adults had less than basic literacy skills. That's about one in seven.

Even if a person does have basic literacy skills, he or she may still not be able to read complicated procedures or understand high-level language. Knowing your users can help you choose the appropriate language level to reach them. How do you gauge an appropriate reading level for a general-public audience? Many companies aim for a fifth- to

---

* These laws have exceptions, the most notable being for children operating farm equipment on the family farm.

eighth-grade reading level, depending on the breadth of the audience. To determine what reading level a sample of writing requires, you can apply any of several readability indices, such as the Gunning Fog Index and the Flesch Readability Index. Calculators are available online.[15] Be aware, however, that all of these indices merely measure sentence length and number of syllables per word. In other words, they are based on the idea that short sentences and short words are easier to read than long sentences and long words. While that may be generally true, these scales do not measure many other aspects of writing that also contribute to readability and comprehension, such as coherent organization, smooth transitions, and so on. They are a help, but not a guarantee.

ARE YOUR USERS SOPHISTICATED?

No, you don't need to know whether your users drink fine champagne or wear the latest Paris fashions—"sophisticated" in this context refers to a concept in products liability law called "the sophisticated user." The idea behind it is that a person who is a professional in a particular area can be assumed to have certain knowledge—and the manual writer does not need to restate things that this sophisticated user is sure to know simply because he or she is in that business. For example, a licensed electrician would be considered a sophisticated user of electrical equipment. A physicist would be considered a sophisticated user of a mass spectrometer. And a fire fighter would be considered a sophisticated user of a self-contained breathing apparatus (SCBA). When you write a manual for a sophisticated user, you can use technical terms you could not use for a general-public reader (although you should still limit your use of jargon), and you can eliminate some basic background information.

A problem arises when you cannot be certain that your reader is a sophisticated user, even though that is the intended market segment. For example, heavy construction machinery or specialized tools may be intended for use by experienced construction professionals—but they may also be available on the rental market. The intended user for a jackhammer may be a career concrete worker, but nothing prevents the weekend do-it-yourselfer from renting one to remove an unwanted concrete patio in his backyard.

Quite often, the same product will have two sets of users. In that case, the product may have two manuals, written at different language levels, or one manual with separate manual sections, some written for professionals and some written for general-public users. In the second case, each section must be clearly labeled so users know whether to read or skip.

Example 2.3 contains two passages that describe the same mechanism. The writer has altered the style, however, to meet specific users' needs. These two passages could occur in the same manual—the first in a section directed to surgeons or trained technical personnel, and the second in a section directed to equipment assistants and non-technical personnel.

Example 2.3: Changing Style for Different Users. (Adapted from Cooper Vision Systems Operators Manual, Model 8000, Cavitron Surgical Systems, Irvine, CA. Used with permission.)

Passage A—For the Professional

A peristaltic pump is used to create the suction for vacuum. You can use different vacuum levels for the various handpieces and "tips" employed during surgery.

A constant volume peristaltic pump delivers a constant flow rate of 28±2 cc/min. The pump is driven by a regulated DC control voltage that has less than 2% output variation of 106 to 128 VRMS and a load variation of 0 to 75 ounce-inches of torque.

Passage B—For the General Public User

Venting action causes the vacuum level at the port to drop to less than 2 inches of water in less than 300 milliseconds and occurs automatically each time the foot switch moves from Position 2 to Position 3.

An important part of the vacuum system is the vent control. A simple way to understand venting is to think about what happens to fluid in a soda straw. The fluid is sucked up into the straw. (This is analogous to the vacuum buildup taking place in the ABC system.) If you place your finger over the top of the straw, the vacuum is maintained and the fluid stays in the straw with little or no leakage. To release the fluid, remove your finger from the top of the straw and allow air to enter. In essence, this is what happens in the vacuum system of the ABC machine.

Once you have identified your users, the next task is to conduct a hazard analysis on your product.

WHAT ARE THE HAZARDS AND WHAT SHOULD WE DO ABOUT THEM?

When you know who is using your product, you have half the information you need to keep them safe. The other half is figuring out how they could get hurt (or other damage could result). For that you need to know the product's hazards. A hazard is a source of potential injury or damage associated with a product. Examples of hazards include the sharp blade of a meat slicer, an ingoing nip between two rollers on a printing press, a high-voltage electrical connection, and so on. Hazards can also involve the interaction of one product with another—such as mixing a drain cleaner containing hydrochloric acid with bleach (sodium hypochlorite). The two react to produce chlorine gas, which at high concentrations is deadly.

What Is a Hazard Analysis?

Essentially, a hazard analysis identifies ways in which a user (or someone else) could become injured or damage could result in connection with a product. Do not confuse hazard analysis with failure analyses, such as Failure Modes and Effects Analysis (FMEA). While the two are related, FMEA typically focuses on how product components might fail and what the results would be. For example, an FMEA of a household gas furnace might consider these potential failures among others:

- Power failure (electric ignition won't function to light furnace)
- Electric ignition switch failure (electric ignition won't function to light furnace)
- Gas leak (insufficient pressure to ignite; potential for explosion or fire)
- Plugged flue pipe (carbon monoxide diverted into living space)

These all relate to the failure of the product to work as intended. A hazard analysis (for our purposes) relates not just to abnormal operation, but also to adverse events that could occur *when the product is operating normally.* Of course, you also need to consider (and warn about) hazards with abnormal operation, such as warning against using an electrical product with a damaged cord, but often the greater challenge is identifying all the hazards present in normal circumstances.

You may also want to conduct a *risk analysis.* The hazard analysis identifies the hazards that could produce injury or damage, and the risk analysis estimates the likelihood that an adverse event would occur. To illustrate the difference, consider a modern walk-behind gasoline-powered lawnmower with a "dead man switch" that stops the blade from turning when the operator's hand leaves the handle. The *hazard* is the spinning blade that could amputate fingers if the operator reached into the discharge area to clear a blockage. The *risk* is the likelihood of that event happening. If the dead man switch has not been disabled, the risk is very low, since it would be nearly impossible to hold the handle (preventing the dead man switch from engaging) and simultaneously reach into the discharge chute. However, the risk might be considerably elevated if the dead man switch were easy to disable and if the operator were motivated to do so (for example, because the switch not only stopped the blade, but also shut off the engine, requiring a restart each time).

Why Do a Hazard Analysis—And When Should We Do It?

Is it necessary to do a formal hazard analysis? After all, you know your product, right? Do you really need to go to the trouble of a formal process to find out what you already know? A formal hazard analysis is an important step for three reasons:

- Your product may have hazards that are not obvious or easily recognized.
- A hazard analysis will help you make good decisions about how best to mitigate hazards found.
- In a products liability lawsuit, having a systematic hazard analysis process in place shows that your company is making a good faith effort to keep product users safe.

The answer to when to do a hazard analysis is like the old joke about when to vote in a rigged election: early and often. Hazard analysis is never a one-time operation, but should be repeated as needed.

New Product Development

The best time to start a hazard analysis is when the product is starting to be developed. If the product is early in the design phase, a hazard analysis may identify hazards that can be solved with a simple tweak to the design. If the product is in the form of a CAD drawing, it's a lot easier and cheaper to make that design change than if the product is already in production. But how will you know that a new product is being developed? The best way is to start showing up at new product development meetings, even if you haven't been specifically invited. They probably won't throw you out, and after you've been to a few sessions, your presence will be expected.

Existing Products

If there has been no formal hazard analysis for a product that is already being made and sold, now is the time to do one. The ultimate goal is to have in place a systematic and consistent process for identifying and mitigating product hazards across all product lines. If your company manufactures many different products, implementing such a comprehensive program will take time, possibly many years. But just because it cannot be done all at once, don't give up. Pick a specific product or product family, and start there. Much of the time and effort in a hazard analysis goes into planning and developing the process. Once that is complete for one product or product line, it is easily adaptable to others.

Model Changes

Even if a product has had an initial hazard analysis, it's a good idea to revisit it whenever there is a model change. Most companies strive to keep improving their products to gain or keep a competitive edge. "Continuous improvement" is more than a buzz phrase—it's a corporate mindset. The result is that products change, and when they change, sometimes old hazards are removed and new ones introduced. If the change is minor, checking that the most recent hazard analysis is still applicable may be a very quick process. On the other hand, a major change may require a whole new analysis.

Throughout the Life Cycle of the Product

If the product is an industrial product with a long expected useful life, it's a good idea to review (and sometimes redo) the hazard analysis periodically. Even if the product itself doesn't change, the user group may change. For example, industrial products with various hazards relating to moving parts (gears, belts, etc.) may have been operated with few or no associated injuries for many years, but then suddenly accidents start to happen. What's changed? Not the machine, but the people using it.

For decades, the machine may have been operated and maintained by people who had worked with machinery of one sort or another since childhood. If at age 12, you get your finger pinched between the chain and sprocket of a bicycle as you are lubricating the chain, you learn to keep your hands away from moving parts. As an adult working in industry, that early pinched finger might help you recognize the dangers and avoid injury. But if instead of tinkering with a bicycle at age 12, you spent your free time playing video games, you would not have learned the same lesson (although you may have developed other skills, such as hand-eye coordination). Similarly, if you are accustomed to products with many built-in safety features, especially passive safety features (like a dead man switch), you may not appreciate just how dangerous some older machinery can be.

Experience base is not the only factor that can affect safety. Ergonomics also plays a role. Perhaps when that machine was designed and for many years thereafter, the expected operator was a male born in the United States. The average height for U.S. men is about 5'9". But now, because of changing demographics and immigration patterns, perhaps the more common operator is a female immigrant from Vietnam. The average height for Vietnamese women is just over 5'. That size difference typically also means differences in arm and leg length, which may affect the ability to reach hand controls or pedals. A machine that was designed to be reasonably safe when operated by someone 5'9" tall may be significantly less safe when the operator is a good deal shorter.

Periodically reviewing a hazard analysis for long-lived products, particularly when you know that the user group has changed, is essential. While the product does not normally suddenly become defective just because the user group has changed, you may find that changes to the warnings and instructions are needed. In some cases, the company may decide to offer retrofit options to make the product safer (pedal extensions, for example).

HOW TO CONDUCT A HAZARD ANALYSIS

Many different methods have been developed to identify hazards and estimate risk, but regardless of the method chosen, you will want to make sure that your process is both *systematic* and *comprehensive*.

Make It Systematic and Comprehensive ·

A systematic procedure means simply that you follow a well-thought-out plan when conducting the hazard analysis—and that you follow the same procedure every time, for every product. A good plan should address at least the following aspects:

- Point(s) in product development/life cycle for conducting a hazard analysis
- Personnel to be involved
- Procedure for identifying the expected user group(s)
- User characteristics to be evaluated
- Conditions of use to be evaluated
- Types of hazards to be addressed
- Procedure for ranking hazards according to severity and likelihood
- Procedure for selecting mitigation options

A comprehensive plan means that you consider each element in your analysis as broadly as you reasonably can. If your expected user group is adults working in a manufacturing plant, is there a possibility that the product would be used in another setting? Or that the users might not be adults (in a country without laws against child labor)? Do some hazards only appear in extreme environmental conditions? If normal practice is for your product to be installed by factory-trained professionals, might it instead be installed as a do-it-yourself project by the purchaser or even by the receiving company's own maintenance staff? Your goal, to the degree practicable, is to generate a comprehensive list of potential hazards that covers all situations that could occur.

You may find that some hazards only emerge in certain combinations of circumstances. Does that mean that you have to warn about or otherwise mitigate every conceivable hazard, no matter how remote or far-fetched the hazard and circumstance? Not at all—but casting a wide net with regard to users and circumstances of use as you analyze the product makes it much less likely that you will miss a hazard that might be reasonably likely to occur, but just not every time. An added benefit is that if a lawsuit does occur, you can show that you did your due diligence in attempting to identify and mitigate the hazards associated with your product.

The analysis should address not only the product itself, but also accompanying literature, including marketing material. Make sure that marketing materials do not conflict with what the manual says about how the product should be used. And make sure the manual is in concert with the product design. Of course, if you find problems with the manual, redesign of the manual is the only option—you can't very well put a warning label on the front cover saying, "WARNING! This manual may be hard to understand."

What Method Is Best?

Many different approaches have been developed for identifying hazards and estimating risk. A partial list includes these:

- Preliminary Hazard Analysis
- Failure Modes and Effects Analysis
- Event Tree and Decision Tree Analysis
- Fault Tree Analysis
- Probabilistic Risk Assessment

In-depth discussion of these is well beyond the scope of this book. A good resource for an overview of various methods is *Risk Assessment* by Ostrom and Wilhelmsen.[16] Some of these methods are quite complex and use statistics and mathematical probabilities to estimate risk. While some larger companies may employ in-house personnel equipped to do this sort of analysis (or have the resources to outsource it), such complex methods may be out of reach for smaller companies. The good news is that usually you do not need to undertake such a complex analysis.

Most of these methods simply offer a rigorous, step-by-step logical framework to use to look for hazards. For example, a fault tree analysis starts with an accident and works backward, using logic gates (AND/OR), to find failures (either of components like electrical switches, or processes like hand-washing in a restaurant) that led to the ultimate problem. Chances are that for your products, you (and your colleagues) would be able to do something similar.

Suppose your company manufactures a motorized commercial meat slicer that cuts bologna into thin slices, which are then packaged for sale. The large tubes of bologna ride on a conveyor belt through the machine where a spinning blade slices the product. You know that one of the challenges inherent in food processing equipment is that the equipment needs to be cleaned regularly. Whenever you have a sharp blade and someone needs to clean it, there is the potential for an accident. Think about what circumstances would have to occur for someone to get seriously hurt. These might include

- A guard being removed for cleaning AND
- The machine being energized during cleaning

If either of these is excluded, the potential for serious injury is reduced. If the guard is not removed, it would prevent a hand from contacting the blade. If the machine is not energized, the worker might still cut his or her hand on the stationary blade, but would be much less likely to suffer an amputation. You can go through a similar exercise for each potential hazard that you have identified, analyzing the actions and circumstances that must take place for an accident to occur.

Estimating risk is more challenging. It's relatively easy to determine the probabilities for failure of product parts, like springs or switches or belts or sensors. These components are designed with an expected useful life and failure rate as part of the design calculus. You can plug these numbers in and estimate the likelihood of a failure at any point in the product's life cycle. Human action is more difficult to predict. Turning off the machine before cleaning it is an either-or proposition, to be sure, but the probability is not 50–50 like a heads-or-tails coin toss, because the decision is not random. That is why your hazard analysis must consider users and conditions of use as well as the product itself.

Some of the most effective hazard analysis systems are set up as a series of questions, with checkboxes or open-ended response options. One simple approach is to create a hazard analysis form, such as the generic one shown partially in Example 2.4. Tailor it to fit your product line, and then use it to guide your analysis. Once the form is complete (and the more detailed the responses, the better), you can use the information to classify and prioritize the hazards to help you identify an appropriate response to each one.

Example 2.4: Sample Hazard Analysis Form

PRODUCT INFORMATION

Name/Description: _____

_____

Model #: _____

New Product? Yes _____ No _____ Previous Model #: _____

Development Stage: _____

_____

CONDITIONS OF USE

Location: Indoors _____ Outdoors _____ Both _____

Temperature range: _____ Humidity range: _____

Environment: Clean ___ Dirty ___ Dusty ___ Wet ___ Windy ___ Rain/Snow ___

Special or extreme conditions: _____

_____

EXPECTED USERS

Age range: _____ Sex: M _____ F _____

Height range: _____ Weight range: _____

Education: _____ Special training or certification required? Yes ___ No ___

Previous experience with similar products? Yes _____ No _____

User-maintained? Yes _____ No _____

HAZARDS
Mechanical
   1. Describe hazard (e.g., rotating gears): _____

       _____

      a. When is it present? _____

      b. Proximity to operator or service person? _____

      c. Redesign feasible? Yes _____ No _____ Explain: _____

       _____

      d. Possible mitigation options (e.g., guard, lockout/tagout): _____

      e. Nature and severity of hazard (e.g., amputation of arm): _____

      f. Mitigation plan. (For each hazard identified, describe planned mitigation, i.e., on-product label, design change, safety message in manual, etc. Include rationale for mitigation chosen.) _____

      h. Mitigation achieved: Date: _____ Initials: _____

   2. Describe hazard (next one—answer a—g for each hazard)

Electrical
     1. _____

     2. _____

Chemical/Toxicity
     1. _____

     2. _____

Burn
     1. _____

     2. _____

Other
     1. _____

     2. _____

Are There Regulations and Standards That Apply to Hazard Analysis?

Depending on what industry your company is in, there may be both regulations and standards regarding hazard analysis. For example, if your manufacturing process uses hazardous chemicals, you may be required to perform a chemical process hazard analysis under the Occupational Safety and Health Administration (OSHA) Rule for Process Safety Management of Highly Hazardous Chemicals (29 CFR 1910.119). If so, the Department of Energy has published a handbook to assist you in designing and conducting a hazard analysis.[17] The handbook is worth looking at even if your product has nothing to do with hazardous chemicals, as it shows step-by-step how to develop and conduct a hazard analysis involving a process. Chances are that if your company is required to comply with one or more government regulations, management is already well aware of the fact.

If you are in an industry without specific regulatory requirements, some more general resources are available that may be helpful. The American National Standards

Institute (ANSI) publishes the B11 series of standards relating to machine safety, and also publishes a series of standards relating to risk management and risk assessment. These are the ANSI Z690 series, of which some are shared with ISO 31000 (ISO is the International Organization for Standardization). Finally, OSHA offers several tools related to job hazard analysis (since OSHA is concerned with worker safety). Some of these may be useful to help identify hazards that users of your products might encounter. Search "OSHA hazard analysis" to find a variety of resources.

SHOULDN'T THE ENGINEERS BE DOING ALL THIS?

You may be thinking, "Hey, I'm a technical writer, not a design engineer or even a safety engineer. Why would I need to be concerned about hazard analysis?" The answer is that hazard analysis is not a one-person job—it's a team sport. The composition of the team may well determine how effective the hazard analysis is, and tech pubs should be part of that team.

Why You Need a Cross-Disciplinary Team

It might seem as if the simplest approach would be to ask the design engineer to identify the hazards in the product. After all, no one knows more about how the product is designed to function than the person who created that design. That may be true, but it's also the reason that the design engineer should not be the only one involved. The designer of a product knows too much about the product. He or she would have a very difficult time looking at the product as a first-time user would. Questions that a first-time user might have that the manual doesn't address (what happens if the wi-fi signal extender is subjected to freezing temperatures during the winter in my unheated summer cottage in the north woods?) simply wouldn't occur to the design engineer—either because he or she already knows the answer, or more commonly, because it wouldn't occur to the engineer that the situation would arise.

The best approach is to assemble a cross-disciplinary team representing a variety of perspectives in the company, such as engineering, service, sales, production, marketing, legal, and technical publications. That way you have the benefit of different knowledge bases and viewpoints. A hazard analysis by a cross-disciplinary team that included field reps might have averted the accident involving the grain bin referred to earlier in this chapter, since the sales reps were aware of a hazard the design engineer never considered.

"Shop-Blindness" Can Be Costly

Shop-blindness refers to the tendency for people working in a particular industry or with a certain product to assume too much knowledge on the part of the user. Shop-blindness can cause you to overestimate your users' knowledge of product hazards, resulting in a missed hazard that could be addressed with a warning. Failing to include a warning always to put two new tires on the rear wheels might become an expensive issue in a lawsuit resulting from an oversteering crash.

How can you guard against shop-blindness when conducting a hazard analysis? That's one of the benefits of a cross-disciplinary team—even an in-house marketing expert in the power-tool company may not know that a hand-held wood router can only be safely moved in one direction along an edge. Another option is to include someone outside the manufacturing arena entirely.

The "Person on the Street" May Be Your New Best Friend

You may find it helpful to include someone on your hazard analysis team from outside the company or at least outside the product world. For example, you might borrow an employee who works in finance. That person may know all there is to know about balance sheets and depreciating capital assets, but know next to nothing about the product your company manufactures. Often the novice will notice potential hazards that those experienced with the product miss, simply because the novice doesn't know how the product is *supposed* to be used. A novice perspective can be extremely valuable in identifying hazards and pointing out possible ways that a first-time user might go wrong.

WE FOUND A HAZARD. WHAT NOW?

Chances are, you found more than one—maybe a whole list of them. Even if your list is fairly long, it doesn't mean that your product is unreasonably dangerous. Not all hazards are created equal. Follow this process to sort them out and address them:

1. For each hazard, assess *severity* and *likelihood*.
2. Apply the Hazard Control Hierarchy (explained below) to each hazard.
3. For residual hazards that cannot be otherwise eliminated or controlled, develop warnings and/or instructions.

Assess Severity and Likelihood

For each hazard identified, consider what would be the most severe possible consequence if someone interacted with that hazard. For example, a hand coming in contact with a table saw blade could result in amputated fingers—but could also result in a simple cut. For this exercise, the most severe consequence would be an amputation. How should you categorize the various possible severities? If you conduct an Internet search for hazard control, you will find many systems. Some rank the severity using numbers (e.g., 1–5, with 5 the most severe); others use words (e.g., "negligible" to "catastrophic"). If your company already has a system in place, use that. Otherwise, you can use one of the many available on the Internet or design your own. Whatever you choose to do, keep two principles in mind:

- Make it simple rather than complex.
- Define what each level means.

A simple system using three to five levels of severity is much easier to manage than a complex one with eight or ten levels. Too many levels and the team may get bogged down in discussing whether a particular injury should be assigned to Level 7 or Level 8, but if you only have three levels, it's much easier to come to agreement. Whatever number of levels you end up with, agree on a definition for each level. For example, here is a simple three-level system with definitions:

1. Low: Minor injury, requiring no more than on-site first aid with no work-time lost; Minor damage that does not interfere with proper equipment functioning and results in no downtime.

2. Medium: Moderate injury, requiring skilled medical care, such as cuts requiring stitches or bone fractures, resulting in lost work time, but not permanent disability; Moderate damage that requires equipment repair and results in downtime of less than one week.
3. High: Death or major injury requiring hospitalization and possible permanent disability; Major damage requiring replacement or rebuilding and requires downtime of more than one week.

The definitions should be sufficiently clear-cut that it is easy to determine the appropriate level for a given injury or amount of damage.

Once you have divided up your list of hazards by severity level, the next step is to assign a likelihood or probability to each one. This process will necessarily be somewhat subjective, unless you happen to have an exhaustive injury history for your product. For hazards that involve human actions, such as attempting to clear a machinery jam without first locking and tagging out the power to the machine, it may not be possible to determine a precise mathematical probability. As with severity, however, you should be able to agree on a relatively simple probability system, such as again using LOW, MEDIUM, and HIGH. If possible, when determining probability, include people knowledgeable about product use in the field, either customers or field personnel such as technical sales people or service technicians. Their estimates of likelihood will often be based on actual uses they've either witnessed or heard about from others.

The end result of this step is that each hazard you have identified can be located in a simple matrix. For example, suppose your product is the bologna slicer mentioned earlier in the chapter. One hazard that is found during the hazard analysis is that someone cleaning the machine with the guard removed and the machine *not* energized, could sustain a severe laceration to the hand, requiring stitches. The severity is Medium. The probability is determined to be Medium (personnel are supposed to wear cut-resistant gloves when cleaning the machine, but the gloves make the procedure more awkward, so sometimes the procedure is done bare-handed). The matrix would look like this:

Severity

High Severity Low Probability	High Severity Medium Probability	High Severity High Probability
Medium Severity Low Probability	Medium Severity Medium Probability Laceration	Medium Severity High Probability
Low Severity Low Probability	Low Severity Medium Probability	Low Severity High Probability

Probability

FIGURE 2.2   Blade laceration located on hazard matrix.

Apply the Hazard Control Hierarchy

Ideally, it would be possible to design products that worked cost-effectively to achieve their purpose without having any associated hazards. Normally, that is not possible. While it's theoretically possible to design a completely safe automobile, it wouldn't go very fast and wouldn't get very good gas mileage. Because most products have some associated hazards, there is a well-accepted hierarchy of hazard control. The order of preference is:

1. Design the hazard out of the product.
2. Shield or guard the hazard.
3. Warn against the hazard.
4. Instruct about the hazard.
5. Ignore the hazard.

Clearly, the best choice is to design hazards out of the product. If the hazard is no longer there, it cannot cause an injury. Sometimes a hazard can be "designed out" by substituting a safer alternative. An example is an air-circulating fan with soft vinyl blades instead of steel: even if a finger is poked into the fan while it is running, the soft vinyl blades do little harm. Naturally, as noted earlier, designing the hazard away is much easier when done early in product development, which is why it is so important to start the hazard analysis process early.

Sometimes, hazards cannot be designed out. An example might be a printing press, in which the turning rollers produce ingoing nip points, where an operator's hand could be drawn in and crushed. Designing out that nip point would mean that the press could no longer do what it was intended to do. In that case, an alternative would be to provide a guard that shields the nip point while the press is operating. Physically removing the hazard from human interaction, either by a guard or some other barrier, is also referred to as providing *engineering* controls. Providing guards is less desirable than designing out the hazard because guards can be removed. In fact, a frequent question at seminars is, "Are we liable if somebody gets injured when a guard has been removed?" Whether the manufacturer is liable in that situation depends on many factors, but two critical questions to ask are

- Why was the guard removed?
- How easy was it to remove it?

If the guard was removed because having it in place significantly interfered with the operator's ability to run the machine efficiently, then perhaps changing the design would have been a better choice.

If the guard is easily removed (e.g., unhooked and lifted off), it will be more likely to be taken off than if it is difficult to remove (bolted or welded on, requiring tools to remove). If a guard covers an area of the machine that should only be accessed when the machine is not running (and perhaps even locked out), making the guard difficult to remove is probably a good idea. On the other hand, if sometimes the guard needs to be removed—or at least moved—while the machine is in operation, perhaps it should *not* require tools—that might make it less likely to be replaced.

Often manufacturers place warning labels on guards reminding users to that the guards should be in place when the machine is operating. One strategy to consider is also applying a warning label that is visible only when the guard has been removed—informing the user that a guard is missing and that it must be replaced before operating the machine.

If a significant hazard cannot be designed out or guarded against, the manufacturer must provide a warning. Warnings and instructions are called *administrative* controls. Instead of physically separating the hazard from the human, administrative controls depend on the cooperation of the human to avoid injury by following specified procedures for safe operation and/or wearing appropriate personal protective equipment (PPE). Naturally, administrative controls are a less reliable means to prevent injury or damage, since humans do not always behave as we wish they would. For this reason, warnings should be reserved for "residual" hazards—that is, hazards that cannot be designed out or effectively guarded.

Do not try to fix a flawed design by adding a warning. The courts have repeatedly held that a warning is no substitute for good design or adequate guarding. For example, in the case of the meat slicer discussed above, it would *not* have been acceptable for the manufacturer to leave the blade unguarded, but to provide a warning—even a "perfect" warning.

If you decide that a warning is in order, you must next determine whether the warning should be an on-product label or a warning in the manual. How can you decide when to put a warning label on the product itself and when to put the warning in the manual? The answer depends on several factors, including seriousness of the hazard, size of the product (and therefore available space for labels), intended users, and the use environment of the product. A chainsaw, for example, is a relatively dangerous product with serious hazards, including blade kickback, fire, and explosion from the gasoline fuel, eye injury from flying chips, etc. It's also a relatively small product (especially the small "homeowner" models, which are used by the least experienced users), and it becomes very dirty during use. The available space for labels is small and any labels used are likely to be abraded or covered with oil and dirt rather quickly. Another example of a relatively dangerous product with little option for on-product warnings is an airless paint sprayer. As anyone who has ever owned or rented one of these knows, after the first few uses, most of the sprayer gets covered with paint, so on-product labels become quickly obscured.

In deciding what warnings to put directly on the product itself, ask three questions:

- Is this hazard both serious and likely?
- Is this *must-have* information for users?
- Is an on-product label practical?

If the answer to all three is "yes," then an on-product label is certainly in order.

Sometimes the answer to the third question requires creativity. In the case of the airless paint sprayer, for example, rather than use a traditional stick-on label, one company engraved the critical warning (skin injection hazard) on the nozzle—the only part of the product that routinely was carefully cleaned.

Do not confuse *practical* with *attractive*. Sometimes marketing departments will resist putting warning labels on products because they worry that the label will either spoil the aesthetics or make the product seem dangerous. If a label is needed, it must be provided—the consequences of not fulfilling your duty to warn can be far more damaging than a few lost sales.

For hazards that do not require a specific warning, the manual offers another means to get safety information to users. Some hazards may be dealt with adequately within the operating instructions for the product. Recognize, however, that you cannot guarantee that information in the manual will get to the user, so reserve this option for relatively minor hazards that users are not likely to encounter. A later section of this book describes in detail how the manual can be a significant safety resource—and addresses the problem of making sure that the user has access to the manual.

Some very minor hazards can safely be "ignored"—meaning not actively mitigated. But won't that make the company liable if someone is injured? Possibly. But the tradeoff may be worth it. If you choose not to mitigate a hazard that would cause, say, a pinched finger, even if someone suffers that injury and decides to sue you for failure to warn, your attorneys should be able to settle the claim easily for a relatively small amount of money. On the other hand, if you slap a warning label on the product for every minor hazard and the result is that another user is killed or maimed because a serious hazard warning was lost in a sea of labels, your company or its insurance company may have to pay a hefty damage award. Even if you decide to ignore a hazard, document that decision and the rationale behind it.

When you have completed this step of your hazard analysis, you should be able to create a table such as the partial one shown in Table 2.1 for a table saw.

Apply a Rational, Standardized Mitigation Process

When you have completed the hazard analysis, take a second look at the proposed mitigation choices. When possible, similar levels of severity and likelihood should be mitigated in similar ways. For example, if you have two hazards, both assessed as high severity/medium likelihood, but one is to be mitigated with a guard and one with an on-product warning, that difference might raise a red flag—or there may be a good reason for handling them differently.[18] A guard might not be possible if it

---

TABLE 2.1
Partial Hazard Analysis for Table Saw

Hazard	Severity	Likelihood	Mitigation	Comments
Sharp blade— amputation	Medium/high	Medium	Guard Push stick	Guard must be removable for some cutting jobs. Push stick should have convenient storage location on saw table.
Kickback— lacerations, eye injury	Medium	Medium	Anti-kickback pawls Warning to wear safety glasses	Pawls must be removable for some applications. Provide safety glasses with saw?

would interfere with the functioning of the product (you cannot place a guard over the cutting teeth of a chain saw without rendering it useless for cutting). The key is to make sure that the choices made are carefully and deliberately thought through and are as consistent as possible. If someone is injured, the plaintiff's attorney will take a close look at how similar hazards have been handled and may question your decisions. If you have had to use a lesser means of control than the severity and likelihood would seem to warrant, you want to be able to explain the rationale behind that choice.

Similarly, when you are conducting your hazard analysis, it's a good idea also to look at your competitors' products and compare how similar hazards have been addressed. If the same hazard is guarded on a competitor's version of your product, but you plan to address it only with a warning in the manual, you might want to revisit that decision. On the other hand, if your product has the guard and your competitor's has only a warning, you may be ahead of the curve!

Document, Document, Document!

As you work through a hazard analysis for one product and develop a system for determining severity, likelihood, and appropriate mitigation strategies, make sure that the process is recorded. You want to have comprehensive and detailed records of the hazard found, the mitigation strategies chosen, and the reasoning supporting those decisions. The old saying that "if it's not written down, it didn't happen" holds true for hazard analysis. Keeping good records means not only that you will be better prepared to defend against a products liability lawsuit, but it also means that it will be easier to ensure that the process is consistent from product to product.

Technical writers and engineers often express concern about whether such a paper trail can be used against them. While such a record is subject to discovery by the opposing side, ask yourself which would look better to a jury: comprehensive and systematic hazard analysis records or no records at all? And if that doesn't convince you that documentation is a good idea, think about what you will answer when the plaintiff's attorney asks you on the witness stand, "Did you conduct a hazard analysis?" Of course, you'll have to answer "Yes." The next question is sure to be either "Did you document the results?" or "Where are the records of that hazard analysis?" What will you answer then? Unless you have documentation available, you will either look like you are lying about conducting the hazard analysis or like you're trying to hide something.

Your best option is to develop a standard format for formally documenting the hazard analysis process. With a formal reporting system in place, informal notes from brainstorming sessions and problem-solving meetings need not (and should not) be kept as part of the file. As one attorney put it, "Just be sure that the report shows that when you found a problem, you did something about it."

Finally, always follow your company's document retention policy (and if there is no policy in place, try to get one established!). If company policy is to dispose of all records after some period (for example, ten years), follow that policy. The company may designate some records to be kept permanently. If so, make sure you know whether the hazard analysis documentation is considered to be a permanent record or one to be discarded after a period of time. Follow the policy carefully. You want to avoid a situation in which the hazard analysis documentation is retained for one product but discarded for another—that inconsistency could be damaging in a lawsuit.

SUMMARY

Chapter 1 laid out some significant changes that have occurred in the world of instructions and warnings since the last edition of this book. Two of those are a much-expanded user group as companies have increasingly marketed their products worldwide and a greater emphasis on product safety as product liability litigation, and especially allegations of failure to warn, have grown. Because of these changes, two of the first steps in planning product documentation are figuring out who the product users are and what the product's hazard are. This chapter addressed those tasks.

Manufacturers have a duty to provide reasonably safe products and must provide instructions for safe use. Whether a product is reasonably safe depends in part on who is using it and what level of knowledge they bring to the task. Similarly, what information should be included in the manual varies depending on who the users are and how much they already know. A writer's first task is to learn as much as possible about the product's users, including not just demographics but also how best to reach and engage them. People may not like to read manuals, but most of them will do so, providing that the instructions are designed with their needs in mind. Identifying the characteristics of your user groups is crucial to creating documentation that works for them.

Equally important is knowing the product and the hazards associated with its use. The best approach is to develop a systematic procedure to analyze hazards—ideally beginning at the early stages of product development and repeated periodically throughout the life of the product. Having a systematic procedure and documenting the results helps ensure that hazards will not be overlooked and that similar mitigation strategies are used for similar hazards across the company.

Understanding the audience for your manual and understanding the product and its hazards are key to achieving one of the overarching goals of developing instructions and warnings: keeping the user safe.

CHECKLIST: USERS AND HAZARDS

Personal User Characteristics

- How old is the user?
- Is the user literate? At what reading level?
- Will the user understand technical language?
- Will the user understand mechanical drawings, circuit diagrams, charts, graphs?

PRODUCTS FAMILIARITY

- Does the user operate this machine or product almost every day or only once in a while?
- Does the user have special training in this type of product?
- Is the user likely to have used other products like this one?
- Will the user do routine maintenance on the product?
- Will the user repair the product? *Should* the user repair the products, or should it be repaired only by a trained technician?

PRODUCT SAFETY

- Have we conducted a systematic hazard analysis on the product that considered all the potential users and conditions of use?
- Were there representatives from all parts of the company on the hazard analysis team?
- Has each hazard been ranked according to its severity and likelihood of causing harm?
- For each hazard found, have we developed an appropriate mitigation strategy?
- Have we documented the hazard analysis and the steps taken to improve product safety?

NOTES

1. *Restatement of the Law Torts: Products Liability* (St. Paul, MN: American Law Institute Publishers, 1998), Section 2, Comment a, p. 16.
2. Karen Schriver, *Dynamics in Document Design* (New York: John Wiley and Sons, Inc., 1997), p. 224. This eminently readable and interesting book should find a place on every technical writer's bookshelf.
3. Schriver, *Dynamics in Document Design*, p. 165.
4. Franck Ganier, "Factors Affecting the Processing of Procedural Instructions: Implications for Document Design," *IEEE Transactions on Professional Communication* 47, no. 1 (March 2004): 15–26.
5. See for example, Talha Harcar, John E. Spillan, and Orsay Kucukemiroglu, "A Multi-National Study of Family Decision-making," *Multinational Business Review* 13, no. 2 (July 1, 2005): 3–21, www.proquest.com/, accessed January 12, 2008.
6. For these and other population projections, see the U.S. Census website at www.census.gov.
7. Gail Lippincott, "Gray Matter: Where are the Technical Communicators in Research and Design for Aging Audiences?" *IEEE Transactions on Professional Communication* 47, no. 3 (September 2004): 157–170.
8. F. M. Van Horen, C. Jansen, A. Maes, and L. G. M. Noordman, "Manuals for the Elderly: Which Information Cannot be Missed?" *Journal* of *Technical Writing* and *Communication*, 31 (2001): 415–431 (cited in Lippincott).
9. James H. Melton Jr., "Lost in Translation: Professional Communication Competencies in Global Training Contexts," *IEEE Transactions on Professional Communication* 51, no. 2 (2008): 198–214.
10. Barry Thatcher, "Intercultural Rhetoric, Technology Transfer, and Writing in U.S.–Mexico Border *maquilas*," *Technical Communication Quarterly* 15, no. 3 (2006): 385–405.

11. Quan Li, Menno D.T. deJong, and Joyce Karreman, "Cultural Differences and User Instructions: Effects of a Culturally Adapted Manual Structure on Western and Chinese Users," *Technical Communication*, 62, no. 3 (August 2015): 163–182. See also Alison Gopnik, "Cultural Differences Start Early," *Wall Street Journal Weekend*, July 13–14, 2019, p. C4.

12. Mary Lou Fisk, "People, Proxemics, and Possibilities for Technical Writing," *IEEE Transactions on Professional Communication* 35, no. 3 (1992): 176–182.

13. From a September 2017 UNESCO Fact Sheet, http://uis.unesco.org/sites/default/files/documents/fs45-literacy-rates-continue-rise-generation-to-next-en-2017.pdf. See also *Adult and Youth Literacy: National, Regional, and Global Trends, 1985–2015*, UNESCO Institute for Statistics, June 2013, http://uis.unesco.org/sites/default/files/documents/adult-and-youth-literacy-national-regional-and-global-trends-1985-2015-en_0.pdf, accessed November 29, 2018.

14. The most recent information available that estimates illiteracy rate.

15. See for example, this website: www.readabilityformulas.com/free-readability-formula-tests.php, accessed November 29, 2018.

16. Lee T. Ostrom and Cheryl A. Wilhelmsen, *Risk Assessment: Tools, Techniques, and Their Applications* (Hoboken, NJ: John Wiley and Sons, 2012).

17. DOES Handbook Chemical Process Hazards Analysis, DOE-HDBK-1100-2004, August 2004, www.standards.doe.gov/standards-documents/1100/1100-BHdbk-2004/@@images/file, accessed November 20, 2018.

18. Note that federal regulations cover various machine guarding requirements under 29 CFR 1910. See www.osha.gov/SLTC/machineguarding/standards.html, accessed November 27, 2018.

Part II

The Making of a Manual (Or More Than One!)

3 Designing a User-Friendly Manual

OVERVIEW

Now that you know the functions your manuals fulfill (Chapter 1) and who your users are (Chapter 2), you are ready to begin the design process. Any writer has two major tasks: decide what information to include and decide how to present that information. This chapter looks at what manual users want a manual to do—and what they *don't* want—and identifies strategies for selecting and arranging content to meet those goals. Content is only half the battle: you also need to make a manual that is inviting, engaging, and user-friendly, and this chapter describes effective writing strategies that make it easy for readers to get the information they need.

WHAT DO USERS WANT?

Users want clear information that will help them solve their problems or answer their questions as they use a product, but they don't want to read manuals. And even when they do read a manual, they read as little of it as possible. Interestingly, informal, non-scientific research conducted by the author at numerous seminars over the last thirty years confirms that this aversion to reading manuals holds true even for the vast majority of technical writers *whose job is to write manuals.*

WHEN ALL ELSE FAILS . . .

We've all heard the line, "When all else fails, read the manual." Why not? What is so aversive about reading a manual? Like used-car salesmen and lawyers, manuals have a bad rap. People expect manuals to fulfill one or more of these stereotypes:

- They're boring.
- They're hard to read.
- They're confusing.
- They don't answer my questions.

Where do these come from? The flippant answer, of course, is "experience," but let's look for the germ of truth present in most stereotypes.

They're Boring

Manuals are generally perceived as boring reading. In fact, standard advice for people with insomnia is to get up and spend 20 minutes or so reading a manual until they feel sleepy. What do we really mean when we say something is "boring"? Psychologists define boredom as a *lack of engagement* that is accompanied by an *inability to pay attention.*[1] Instead of asking what makes something boring, let's turn the question around: what makes something *engaging*? Several factors contribute to generating engagement and attention:

- Perceived value of the topic
- Information presented is neither too basic nor too advanced for the learner's level
- Perceived potential for mastery of the information

Information engages people if they see a need to learn it, if it's pitched at the right level, and (most important), if they believe they *can* learn it.

The first of these is intrinsic in user manuals. The value of reading the manual should be learning to use the product. Few of us read product manuals for products we don't plan to use (the author being an exception!). The second is trickier to achieve and depends on good user analysis. Of course, most products have a range of users, so finding a way to meet all their needs is sometimes a challenge. As we shall see, the answer lies in giving different users different paths through the material, so that an experienced user does not need to wade through information he or she already knows. The third is critical for success: unless your user believes mastery of the information is possible, he or she won't see any reason to try. All too often, we discourage readers before they even start.

They're Hard to Read

Have you ever picked up a document and thought to yourself, "This looks hard to read," even before you had made your way through a single sentence? Sometimes the visual impact of a manual is negative—small fonts, not much white space, few illustrations, densely packed pages—all contribute to a first impression that signals a tough go of it to get through the material. Other times an inviting appearance may mask incomprehensible (dis)organization, convoluted, clumsy or ungrammatical sentences, and baffling visuals. Good document design and good choices about content and organization can ensure that manuals meet readers' needs in a user-friendly way.

They're Confusing

We find things to be confusing when we can't follow what's being said or written, either because material is presented out of order or because we don't have sufficient background to understand what's being presented. If a manual directs us to "turn the bevel lock lever to the right" before explaining what the bevel lock lever is and what it is for, we will surely be confused. If a manual directs us to "remove the outer bearing race," and then three steps later in the procedure tells us "before removing the outer bearing race be prepared to catch any ball bearings that fall out," we will not only be confused, we will also be annoyed as we are chasing down escaping ball bearings before they can roll away. Good organization means thinking through the order in which users need information and ensuring that we have not only structured it logically, but that we have also built the necessary foundation for each step. How much foundation is needed depends on how much your users already know.

They Don't Answer My Questions

The last chapter emphasized the importance of avoiding "shop-blindness" and over-estimating your users' knowledge and experience If you have followed the advice in Chapter 2, you should have a pretty good idea of who your user groups are, which allows you to anticipate the questions they will want the manual to answer—providing you can put yourself in the beginner's mindset. Nothing creates frustration faster than having a critical question go unanswered, no matter how many times you page through the manual.

THE UNWRITTEN GOAL OF A SUCCESSFUL MANUAL

This section began with the question, "What do users want?" One thing that users universally want is *not to feel stupid*. Interview after interview with people during usability testing reveals this fact: people feel that a great many products, particularly technology-based ones and their accompanying manuals, make them feel stupid. Technology design visionary Alan Cooper believes that "not feel stupid" is one of users' "personal goals" for using a product/manual. Cooper goes on to say, these "personal goals are always true and operate to varying extents for everyone. Personal goals always take precedence over any other goal, although—precisely because they are personal—they are rarely discussed. . . . Any system that violates personal goals will ultimately fail, regardless of how well it achieves other goals."[2]

When users can't figure out a product and when they feel stupid while trying to read the manual, they have two choices, both of them undesirable:

- They can blame the company and the writers who produced the manual.
- They can blame themselves.

If they blame the company and the writers, it will be that much harder to sell another product to that user. If they blame themselves, they will avoid reading the manual if at all possible, because no one enjoys feeling stupid. Research shows that users tend to blame themselves—even when the problem is bad product design or a poor manual. Schriver found that "users blamed themselves for the problems they experienced more than half the time."[3]

It's not the manuals alone that cause people to feel incompetent. Cooper believes needlessly-complicated products are also at fault.[4] As our technological abilities and our consumer demands increase, and as our product development time often decreases, we experience a growing number of effects resulting from poor design, including these:

- Creeping featurism
- Cognitive friction
- Productivity paradox

Creeping Featurism

Creeping featurism is "the tendency to add to the number of functions that a device can perform, often extending the number beyond all reason."[5] This tendency has become so pronounced, especially in consumer electronics, that some companies are banking on a consumer backlash. A good example is the Jitterbug® smartphone that proudly claims to be the "simplest smartphone ever."

Cognitive Friction

Cognitive friction is "the resistance encountered by a human intellect when it engages with a complex system of rules that change as the problem permutes."[6] A lawn mower or a Swiss army knife is low in cognitive friction; a cell phone or a personal computer is high. For example, the corkscrew on a Swiss army knife remains a corkscrew regardless of whether I am using the knife blade to cut string or cheese (or string cheese, for that matter), but the SHIFT key on a computer keyboard has many different functions, from selecting a block of text to modifying function keys to causing a page break (when pressed simultaneously with another key), among others.

Productivity Paradox

Schriver defines the productivity paradox as the promise of "new technology that is designed to make us work faster and smarter, but actually only makes us work slower and with less confidence";[7] this paradox is perhaps explained by cognitive friction, and in turn explains why many people would rather stick to the old product which they know how to use than to switch to the newer model which they'd then have to relearn.

Good manuals can often mitigate the bad effects of poor design. Exceptionally clear writing can make even an unnecessarily complicated product understandable. Your goals as a technical writer include the overt ones of providing proper information for use and care of the product, conveying important safety information, and so on. But you also have an unwritten goal: make the user feel smart. This is not just good customer relations—it's essential to getting the user to read the manual and learn the product. Remember, before a person can learn new information, he or she has to believe it's possible to master it. Good manuals can help instill that confidence.

HOW DO PEOPLE USE MANUALS?

Products must be designed not only for specific users, but also for the circumstances and manner in which they will be used. We have all seen too often the results when that simple rule is ignored: the umbrella that flips inside-out the first time a stiff wind catches it; the electric coffee pot that spills coffee all over the counter if you try to pour more than a trickle; the warning sign that fades to illegibility when exposed to sunlight. The same is true for manuals.

"Circumstances and manner of use" encompass not only physical aspects, like the environment in which the manual is used (e.g., indoors vs. outdoors), but also functional aspects—how people interact with manuals and what they need to learn from them. Manuals often serve more than one purpose, but even when we focus on the single purpose of providing instructions for safe and proper use of the product, we find that different users may interact differently with the manual. Understanding our audiences and their needs is the key to designing a manual to serve them.

How Do I . . .?

All manuals are by definition "how-to" books. Manual readers want to know how to use the product. They may need to know certain other types of information as well, but the focus of the manual must remain on answering user questions that begin, "How do I . . .?" The specific questions, of course, vary with the product. For an office paper shredder, one question may be "How do I make sure it doesn't jam?" For an industrial stamping machine, one question might be "How do I program it for different thicknesses of metal?" It seems kind of obvious, but it's surprising how often manual writers go astray.

Users want to know not only how to do things, but also how to know when they've done them correctly. Because of their familiarity with the product, writers often take this piece for granted. Software designers have begun to incorporate messages that give the user feedback. Think about the last time you installed a new application on your smartphone. More than likely, when the installation finished, you received a visual indication that the installation was complete: either a message popped up on the screen or a graphic showed the progress. The equivalent statement in a manual dealing with a physical product is something like these:

- Push sensor in until you feel it click into place. When properly installed, only the blue tip should protrude from the housing.
- Tighten belt by adjusting the position of the rear pulley. When properly adjusted, you should be able to deflect the belt about ½ inch by pushing on it midway between the pulleys.
- Turn the adjusting screw by quarter-turn increments until the zero on the gauge exactly lines up with the arrow on the bezel.

Most users approach learning a new product with some anxiety (remember, they're accustomed to new products and manuals making them feel stupid). Including checkpoints that confirm that they have done a procedure correctly (unless it is intrinsically obvious) can help build confidence, reduce anxiety, and improve learning and retention of future material.

What's This Thing For?

In addition to knowing how to use the product, users want to know what the major physical parts of a product are and how they contribute to the product's operation, and they want to know the purpose of every control. When you work with a product every day it's easy to forget than not everyone automatically knows what the "depth-feed module" is or what the "source probe" is or even what the "gravity-return cover" is. Every product has terms that are specific to it. Using them becomes second nature to anyone working for the manufacturer of the product, but they are not intuitively obvious to someone using your company's product for the first time. A user may be very familiar with a similar product made by a different manufacturer, but the same functional part may be called by different name.

If your users are to be able to follow your instructions for how to use the product, they need to know what parts and controls those instructions refer to. Always provide a means to identify parts near the beginning of a manual, usually with one or more photos or illustrations. Then check to make sure that any part or control you refer to later in the instructions is identified on those visuals. Simply learning their way around a product's physical layout will help users to feel more confident and therefore more receptive to instructions.

ALL I WANT TO KNOW IS . . .

No one wants to be forced to read through page after page of information when all that's needed is the answer to a simple question: How do I set the clock on the microwave? How do I adjust the high-pressure limit switch? Where do I put the fabric softener? How do I adjust the infeed rollers for different thicknesses of fabric? On the other hand, all those pages of information may be needed for the first-time user, whether the product is a printing press or a pressure cooker. It turns out that the research supports the notion that users tend to interact with manuals in at least two very different ways.[8]

One group, typically first-time users and those performing a procedure, read the manual *linearly*, that is, they start at the beginning of a section and read through, more or less in order. Despite this linear tendency, hardly anyone reads an entire manual from page one all the way through to the end. But many users, whether beginners or not, often tend to read at least portions of the manual in a linear way. These readers are helped by information that is sequenced meaningfully. These may be step-by-step instructions, explanations of processes, and so forth.

Other users refer to the manual occasionally (and perhaps repeatedly) for specific information, such as the correct spark plug gap for a small engine or the proper lubrication procedure or programming sequence for an industrial machine. These users tend to be more experienced with the product, and their information needs are more precisely focused.

The difference between the two approaches can be likened to the difference between listening to a piece of music and looking up a phone number in a directory. The person listening to music wants to hear it with each note in order. The music makes sense only when heard in sequence. No one jumps directly from the A-flat in Measure 12 to the A-flat in Measure 54. The notes become meaningless when they are played out of sequence. On the other hand, the person looking up a phone number for Mary Smith, hardly wants to have to scroll through the Browns, the Garcias, the Joneses, the Patels, and the Quincys before finally arriving at the Smiths. He doesn't even want to have to read through the Alan Smiths, the Barbara Smiths, and so on. He wants to jump right to Mary—and preferably to the Mary Smith who works in Technical Publications.

To design a manual that will work effectively for all your users, you must understand how they use manuals. A fundamental principle of good design is that "form follows function," and it applies to manual design as much as to product design.

Learn as much as you can about how your users interact with manuals and let it guide your design decisions.

I Don't Have All Day

Most of us would agree that the pace of life has quickened over the last few decades. The availability of email and text messages mean that communications that used to require two or three days to accomplish via postal mail can now take place in just a few minutes. The price for that convenience, of course, is that now we are expected to respond instantly to emails or incoming texts. Attention spans have shortened with the quickening pace, a fact reflected in many ways in our society. For example, teachers are now trained to switch instructional modes (i.e., from lecture to hands-on activity to discussion, etc.) at least every 15 minutes or so or risk losing their students. Television commercials typically switch the image on the screen about once per *second* (some even more often than that) to keep potential buyers engaged.

Understanding the use environment includes recognizing the time pressure that most workers feel. They do not have the luxury of settling back for a leisurely read. Instead, most people out of necessity take the quickest approach they can to get the information they need and avoid reading one word more than necessary.

My Inbox Runneth Over

We live in the information age—but too often that means information overload. For example, do a Google search on "information age," and you will get over *4 billion* results. The only people these days who can truly master the literature of a profession are those who have indescribably narrow fields of study. The rest of us must become experts at searching for the bits of information we need and discarding the rest. A good manual helps that effort by directing readers efficiently to exactly what they need. We read and write many more words per day now than at any time in history,[9] but spend less and less time on any one document. People are used to going through material quickly and selectively. They may well be capable of sitting and reading a lengthy manual, but they are no longer willing to.

The digital age has not only dramatically increased the quantity of available information, but also increased the quantity each of us must deal with every day. Worse yet, it has made it more difficult to delegate some of the tasks of managing it: instead of having a secretary open and sort the mail and reply to routine requests, most of us manage our own electronic inbox. We typically face dozens of emails, many with attached files. Each one requires a management decision before we even start digesting the content: Print it out or read it on screen? File it electronically? Delete it? Reply? All of us tend to spend a large part of our working day just juggling email. Of course, despite the promises of the "paperless" society, we have no shortage of hard-copy documents as well. Most of us have filing cabinets stuffed with paper and stacks of documents on our desks as well.

The point is this: a manual must compete successfully with a lot of other documents if it is to be read at all. Given the sheer volume of information coming at your potential readers, designing a manual becomes an exercise in marketing as well as informing.

You Can't Put a Tractor on a Desk

When we write and design manuals, we typically do so sitting at a computer in a comfortable office. As we ponder how many columns to use and what font size would work best, we are working in ideal conditions. We have good light; we are neither too warm nor too cold; we are sitting in a comfortable chair; and our hands are clean. Depending on the product, the environment in which the manual is used may be very different. If you do not understand the actual physical conditions in which your manual will be used, you cannot make good design decisions for it.

If your product is any kind of industrial machinery, the manual will probably be used wherever the product is located—in a factory or a processing plant or a warehouse. It will probably be noisy, dirty, and have hit-or-miss lighting. The reader will probably not be sitting at a desk, and in fact, may not be sitting at all. It may be cold or hot or steamy. The worker may have to wear some kind of protective gear—goggles or gloves or a respirator. If your product is farm machinery or construction equipment or mining machinery, the manual may be used outdoors in blinding sun or rain or even snow. Or it may be used indoors in a dark machine shed or pole barn—or even underground. The user may have hands that are full of oil or grease or just plain dirt. These less-than-ideal conditions and physical encumbrances make reading a manual difficult at best.

Even if your product is a consumer item for household use, the conditions of manual use may be an issue. Will the user be on hands and knees on the living-room floor examining the underside of the vacuum cleaner? Or craning her neck around the back of a stack of TV receiver, Blu-Ray player, surround-sound amplifier (none of which have enough spare cord length to turn them around) while trying to make out the connection diagram in the manual? Or sitting on the floor of the garage next to a partially disassembled lawn mower engine? Or in the basement trying to squeeze behind the furnace to get at the pilot light for the water heater?

Manuals are almost always written while sitting at a desk. They are almost never read in those ideal conditions.

Make it a point to find out how your manuals are really used. If you can do so, visit sites where your products are in use. Ask user questions about what they are doing and where they are situated when they are using the manual. Ask as many questions as you can think of, including questions like these:

- What's the light like?
- Do you have a place to set the manual down?
- Are you standing, sitting, squatting, kneeling, or lying down?
- Are your hands dirty?
- Do you get interrupted?
- Is it noisy?
- Are you using tools?

The more you know about how your manual is used, the better able you will be to design a manual that works. A manual, after all, is no different from any other tool. If it is going to be useful—and therefore used—it needs to be able to function as intended in the actual use environment.

MANUAL? WHAT MANUAL? KEEPING PRODUCT AND MANUAL TOGETHER

One of the biggest challenges in manual design is finding a way to keep the manual with the product. For industrial products, all too often the manual ends up sitting on a shelf in an office, far away from the machine that is out on the shop floor. For consumer products, manuals either get immediately thrown out, stuffed in a file drawer, or crammed into the kitchen junk drawer. Even if we've kept the manual, the chances of finding it when we need it are pretty slim.

What can you do as a manual designer to address this problem? First, think about what your goal really is. Do you truly want the user to keep the entire manual with the product at all times? Or is the actual goal for the user to have essential safety and operation information at hand when using the product and be able to find reference information when needed? Chances are, the second of these is the real aim. The strategy to accomplish the goal varies with the product. Here are some possible solutions.

- For a large mechanical product like a printing press or a large air compressor, build a compartment into the machine itself for a manual. (Think about your car's owner's manual—it's probably sitting at the bottom of the glove box right now.) You might separate the essential operating and safety information into a shorter manual that would fit in the compartment and leave the reference information in a three-ring binder for the maintenance supervisor's office. You can even chain the manual to the product—but only if the user would not need to take the manual someplace the chain wouldn't reach.
- For a small consumer item like a handheld GPS, create a manual that will fit into the carrying case.
- For portable industrial or consumer products, print essential instructions and reference information on a label applied directly to the product. Be sure the label will be durable under the conditions of use. Alternatively, you could stamp the information into the product housing. Include ready-reference information such as the gas-oil mixture ratio for two-cycle engines or replacement belt size for a vacuum cleaner.
- For machinery of any kind, print a laminated "cheat sheet" of essential information that can be tethered right to the product.
- For all products, put the manufacturer's phone and website right on the product. Make manuals available for download from the website.
- Include a label on the product with a QR code that takes the user to the manual on the manufacturer's website.

Keeping the manual and the product together, like so many other human factors problems, can be solved by a three-step analysis. Ask yourself these questions:

- What is the desired outcome?
- What is currently preventing that outcome?
- What will encourage the desired outcome?

Usually the answer to Question #2 has to do with the use environment. A large manual is cumbersome to take along with a small product. There's no place to put 4-inch three-ring binders on a conveyor belt. The dishwasher manual will clutter the counter or get wet and ruined if it's kept right next to the dishwasher. The more you learn about the environment of use, the more creative you can be in finding ways to make sure your users have the information they need. Like many human factors problems, the key to solving them is to find a way to make it easy for people to do what you want them to. It's easy to keep the car manual in the glove box because it fits and it's not in the way.

By now, if you have followed the strategies in Chapter 2 to identify your audience and its characteristics, and thought through the "circumstances and manner" of use for the manual, you should have a pretty good idea of your target audience and the constraints they face in using your manual. Now it's time to make basic writing decisions: what to include, how to arrange the information, and how best to communicate it.

CHOOSING CONTENT

How do you decide what to include in a manual and what to leave out? The usual advice is to focus on the "need-to-know" information and omit the merely "nice-to-know." But how do you decide which information fits in each category? It's a little like the advice given to a budding sculptor wanting to carve a horse out of a block of marble: "All you have to do is chisel away everything that doesn't look like a horse." Simple it may be; easy it's not.

As we have seen, the research indicates that most users would rather talk to a person than read a manual to become acquainted with a product. One way to think about a manual, then, is as a substitute for a good teacher. Think about the three best teachers you have known. What characteristics did they have in common? Chances are, the best teachers you have known demonstrated the following characteristics:

- They answered your questions—without making you feel stupid for asking them.
- They repeated important information—maybe in different terms or in different contexts—but they made the crucial points more than once.
- If they were teaching skills, like throwing a football or paddling a kayak, they told you step-by-step what to do to accomplish your goal.
- They selected the information they gave you, so as not to overwhelm you with too much, too soon.

These are good guidelines for planning your manual as well.

User Questions as Manual Organizers

Users come to manuals because they have questions they need answered. Some of those questions are specific to the product, but some are general questions common to all products. Step back from your product for a moment and think about the questions first-time users might have. What would they want to know?

- How do I set the copy machine for double-sided copying? How do I make it collate and staple my copies? Will it make color copies?
- This new state-of-the-art injection molding machine looks a lot like our old one to me. What's new? How much retraining will my line operators need to keep it running right?
- How do I program this universal remote for my TV? Will it control the sound bar, too?
- This hydraulic lift gate will make it a lot easier to load and unload our trucks. How do I operate it? Are there any weight limits? What about safety concerns?
- This new washer-dryer combination looks pretty high-tech. Where do I put the fabric softener in?

Imagine the buyers or users of your product talking to your industry vendors or shopping in a dealer showroom. Buyers look at the product, talk to salespeople or vendors, or read the sales literature and instruction manuals. They do this because they have *questions*. A good manual includes content that answers those questions.

YOU CAN SAY THAT AGAIN: USEFUL REDUNDANCY

Did you ever try to stretch the length of a school essay or research paper with a required minimum page length by restating the same information in different words? If so—and if your teacher was on the ball—you probably found the comment "Redundant!" scribbled in red ink along the margin. Or perhaps you find yourself editing out other people's redundant expressions, such as "free gift," or "brown in color," a category of phrases a former colleague[10] refers to as "dog puppies." While redundant expressions can make for bloated writing, not all redundancies are bad. Consider redundant safety systems (a type of "fail-safe" engineering) that ensure that even if the first safety feature malfunctions or is disabled, a second system is in place to prevent injury.

Redundancy in writing can accomplish two aims: to highlight important information and to ensure that readers will see important information. Previews and summaries (such as are used in this book) help the reader to identify and focus on the central concepts. Similarly, repeating particular words or phrases in headings, lead paragraphs, and figure captions will cue the reader that these signal important information. Even more critical in manuals, however, is repeating safety information throughout the manual, wherever it is needed. Such repetition may be redundant from the writer's perspective, but it is not necessarily redundant from the reader's perspective. Remember that users often go to a manual to find specific information to answer a particular question (How do I clear a paper jam? What's the proper spark plug gap? How do I replace the drive chain?), and as a consequence, will read only a very small section of the manual. If all the safety warnings are in a special safety section in the front of the manual—and nowhere else—the user may not see them at all. It makes sense to put safety information wherever it is needed, even if that means repeating it.

HOW TO USE THE PRODUCT IS NOT THE SAME AS HOW IT WORKS

When you were a child learning to ride a bicycle, did your mom or dad explain the principle behind a gyroscope and conservation of angular momentum while running along next to the bike with one hand on the handlebars or behind the seat? Unless your mom or dad happened to be a physicist, probably not. Instead, they most likely urged you to "Keep pedaling!" as you wobbled along on your first few rides. You didn't need to know the physics behind why a bicycle in motion stays upright—you just needed to know what to do to keep it that way.

Manual writers often make the mistake of providing the "why" without the "how to." Just as you do not need to understand how an internal combustion engine works in order to drive a car, you do not need to understand how a product works in order to use it. Not only does excessive technical explanation take up page space and increase manual costs, it can also discourage the reader from using the manual, especially if the explanation is complex. Remember that if material is too advanced for the reader to feel confident of mastery, it is likely to be perceived as "boring."

The tendency for manuals to include too much of "how it works" is especially severe when the product is highly complex or highly technical. With complicated products, the manual writer often must rely on subject-matter experts to develop procedures and instructions. These subject-matter experts frequently are the design engineers for the product—who love to explain how the product works. They're proud of their design, and they want to share with you how nifty it is. Your job as a technical writer and user-advocate is to take that "how-it-works" information and recast it into more user-friendly "how-to-use-it" information.

Some technical information may be needed for the user to make appropriate decisions in using the product. You will need to use your judgment to determine how much theory is appropriate for adequate understanding of your product. User analysis will help you. For example, if you are writing a service manual, you will probably include much more theory-of-operation information than you would in an operator manual. Just remember that the average buyer of a microwave oven, for example, will probably not need an elaborate discussion of the electromagnetic spectrum and the physics of microwaves. He or she only wants to heat a cup of coffee. Focus on the need-to-know.

FOCUS ON THE NEED-TO-KNOW

Remember that a manual is a tool to help the reader do something else—and something physical, at that: laminate plywood, insert a tube of blood into a centrifuge, use a chainsaw to cut wood, or an oxy-acetylene torch to cut metal. Readers of the manual are *users of the product*. They do not read for the act of reading itself. They read to help themselves do something with the product that they can't do without reading the manual. For each piece of information, verbal or visual, you place in the manual, pretend you are a customer of your own product and ask: So what? Why are you telling me this? What's in it for me? Users read because they need. Focus on the need to know. Including non-essential information can overwhelm and confuse the reader—especially the first-time user.

Is there a place in manuals for information that is perhaps not critical, but potentially useful? In other words, where can you put nice-to-know material? Sometimes the answer is, "in another manual." Technology-based products are often accompanied by two manuals: an operator manual and a reference manual. The reference manual may include programming instructions, explanations of the software architecture to enable customization, and other information appropriate for an IT professional. For mechanical products, the technical detail is often to be found in a service manual or parts catalog, which is normally separate from the operator's manual.

An alternative approach is to include a reference section or appendix at the end of the operator's manual. This ensures that all the information is in one book, but arranges it in such a way that different users can easily find what they need.

ORGANIZING CONTENT: COOKBOOKS, NOT NOVELS

It may seem obvious, but manuals are not novels. True, manuals are composed of words on a page, just like a novel, but there the similarity begins and ends. Manuals are not read cover to cover, in a leisurely manner, for pleasure. Manuals are skimmed, read intently only sporadically, and in parts, read with the reader in need and sometimes under stress, and, if vital information is not up front and immediate, perhaps not read at all.

Manuals are more like cookbooks. They are, in fact, read at all only because we can't figure out the product without them. We are also usually in a hurry. After all, we want to record tonight's episode of our favorite series, not read a manual on how to program the smart TV. Or we need to get the new cutters installed to get the line back up and running and meet production quotas. Reading a manual, then, is a means to an end. It is not the end itself. When you create a manual, you must plan for discontinuity, interruption, and stress. Like the writing process, the reading process of a manual is not linear—and not neat.

I WANT TO MAKE AN APPLE PIE—WHY MUST I READ ABOUT BEEF STEW?

Imagine if cookbooks were written with the expectation that they would be read from beginning to end. Some coffee-table-style cookbooks are written with that expectation; typically, these books cover a particular ethnic or national style of cooking (*Exploring Tuscan Cuisine by Mule Train*), are lavishly illustrated with highly saturated photographs, and don't actually have all that many recipes in them. They may get read once and then remain on the coffee table, looking pretty and gathering dust. By contrast, the cookbooks that are actually used to cook—the ones with batter-spattered pages and frayed bindings—are not organized to be read from beginning to end.

These well-thumbed books are structured to make it easy for the cook to find the recipe that's needed. Pie recipes are collected together in a section on desserts. The apple pie recipe is probably even placed together with other fruit pies. These cookbooks get used because they are structured with the user's tasks in mind. Product manuals that get used follow a similar design principle. Major sections of the manual may be aligned with user questions—broad categories of information.

Look at the chart in Example 3.1 to see how a user question can form a section of your manual. The column on the left gives you a typical user question and its corresponding manual section. The columns to the right contain typical answers for two products, a microwave oven and a heavy-duty wrecker. These answers could be used to build a manual.

Example 3.1: Chart Showing Manual Sections Derived from User Questions

Manual Section Derived from User Question	Answer to User Questions about the Product	
	Microwave Oven	Heavy-Duty Wrecker
Application: *What is the main function of this product?*	Used for cooking, reheating, defrosting foods.	Mounted on truck chassis; used for towing, lifting heavy vehicles.
Description of product: *Is this what I'm looking for? Introduce me to it.*	1.7 cubic feet capacity. Turntable. Optional browning element.	Boom and two winches. Two telescoping outriggers on upright mast.
Theory of operation/ intended use: *How does it work? What's it for?*	Microwave cooking in home setting. Not for commercial applications.	Recovery operations using winches; choice of pulling or towing by front or rear wheels.
Special features or design details: *What's special about it?*	Automatic defrosting. Internal temperature-sensor cooking and reheating.	Boom ratings: extended, 12 tons; retracted, 35 tons. Winch ratings: safe load up to 17.5 tons.
Limits of operation: *What can't it do?*	Not to be used with metal or non-microwave-safe plastic dishes.	Ratings apply only if truck chassis is adequate, both winches are attached to load, boom is at least 15° from horizontal, load is lifted vertically.
Setting up/starting: *How do I assemble it? How do I turn it on?*	Plug into grounded three-prong outlet protected by a 20-amp fuse or circuit breaker.	Wrecker installation on truck chassis requires special training.
Normal operation or use: *What is normal use and life of product?*	Cooking/reheating up to two-quart containers of food. Auto-defrost up to five lbs. of meat. One-year warranty.	Heavily dependent on good maintenance and variations in weather conditions. One-year limited warranty.
Turning off/disposal: *How do I stop it? How do I dispose of it?*	Interrupting the cycle. Electrical/computer components require special waste handling and recovery.	Turn-off controls. Emergency operation.

(Continued)

Manual Section Derived from User Question	Answer to User Questions about the Product	
	Microwave Oven	**Heavy-Duty Wrecker**
Abnormal operation: *What tells me something has malfunctioned?*	Arcing, fire. Troubleshooting.	Damage to cable or boom. Binding of cable. Jerky movement of boom.
Preventive maintenance: *How do I take care of it?*	Cleaning. Replacing the turntable.	Complex machine. Separate repair and maintenance manual.
Storage: *How do I store it?*	Protect from freezing.	Putting out of service.
Safety: *How do I use it safely? What should I avoid?*	Safety information will be found throughout the manual and on the product (see Chapter 9).	

Within those major sections, it's best to organize the information around user tasks. Look at the sample manual page for a gas range shown in Figure 3.1. Notice how simply by glancing at the page, you can identify what's covered. In about three seconds flat, you know that this page will tell you how to turn on and adjust the burners. If that's what you're looking for—you've found it. If not, you haven't wasted much time finding out that you need to look elsewhere. Try that with a novel.

MIX AND MATCH: USING MODULAR ORGANIZATION

One of the practices that frustrates product users is receiving a generic manual covering multiple models of products rather than a document specific to their particular model. It's frustrating because as models change, features and physical configurations also change. When the manual says to remove four screws to open the access panel, and the user can only find two screws, he or she is left wondering, "Am I looking at the wrong panel?" Especially for the first-time user, whose anxiety level may already be high, this can add a layer of needless worry. Why do manufacturers do this? Why not publish separate manuals specific to each model? In a word—cost. It's expensive to write and print separate manuals for products with only minor differences among them. The expense is compounded when the manual has to be translated into other languages. Cookbooks face a similar "model change" problem with variations on a recipe type: apple pie, peach pie, cherry pie, etc. They're all fruit pies, and most of the basics are the same for all.

One partial solution is to plan your manual in modular sections that can be assembled into a complete manual. Some portions of the manual apply to all models of a product, maybe even to the entire product line. Others are specific to specific models or products. Cookbooks will often provide a generic recipe for, say, fruit pie, and then list changes for different fruits. If you can design your manual in modular pieces, you can almost custom-make a manual for a product. Here's how it might work.

Setting Surface Controls

CAUTION Do not place plastic items such as salt and pepper shakers, spoon holders or plastic wrappings on top of the range when it is in use. These items could melt or ignite. Potholders, towels or wood spoons could catch fire if placed too close to a flame.

In the event of an electrical power outage, the surface burners can be lit manually. To light a surface burner, hold a lit match to the burner head, then slowly turn the surface control knob to LITE. After burner lights push in and turn knob to desired setting. Use caution when lighting surface burners manually.

OFF ..LITE.. hi 6 5 4 3 2 lo

CORRECT

INCORRECT

Never extend the flame beyond the outer edge of the utensil. A higher flame simply wastes heat and energy, and increases your risk of being burned by the flame.

Setting Surface Controls

Your range may be equipped with different sized surface burners. The ability to heat food quicker and in larger volumes increases as the burner size increases.

The **SIMMER** burner (some models) is best used for simmering delicate food items such as sauces, etc.

The standard burners can be used for most surface cooking needs. Some models include a standard sized center burner (or 5th burner).

The **POWER PLUS** burner or burners (some models) are best used for bringing large quantities of liquid to temperature and when preparing larger quantities of food.

Regardless of size, always select cookware that is suitable for the amount and type of food being prepared. Select a burner and flame size appropriate to the pan. Never allow flames to extend beyond the outer edge of the pan.

Operating the Gas Surface Burners:

1. Place cooking utensil on surface burner.
2. Push the surface control knob in and turn **counterclockwise** out of the OFF position.
3. Release the knob and rotate to the LITE position. Note: All four electronic surface ignitors will spark at the same time. However, only the burner you are turning on will ignite.
3. Visually check that the burner has lit.
4. Push the control knob in and turn **counterclockwise** to the desired flame size. The control knobs do not have to be set at a particular setting. Use the guides and adjust the flame as needed. **DO NOT** cook with the surface control knob in the LITE position. (The electronic ignitor will continue to spark if the knob is left in the LITE position.)

Setting Proper Surface Burner Flame Size

For most cooking, start on the highest control setting and then turn to a lower one to complete the process. Use the recommendations below as a guide for determining proper flame size for various types of cooking. The size and type of utensil used and the amount of food being cooked will influence the setting needed for cooking.

*Flame Size	Type of Cooking
High Flame	Start most foods; bring water to a boil; pan broiling.
Medium Flame	Maintain a slow boil; thicken sauces, gravies; steaming.
Low Flame	Keep foods cooking; poach; stewing.

For deep fat frying, use a thermometer and adjust the surface control knob accordingly. If the fat is too cool, the food will absorb the fat and be greasy. If the fat is too hot, the food will brown so quickly that the center will be undercooked. Do not attempt to deep fat fry too much food at once as the food will neither brown nor cook properly.

*These settings are based on using medium-weight metal or aluminum pans with lids. Settings may vary when using other types of pans. The color of the flame is the key to proper burner adjustment. A good flame is clear, blue and hardly visible in a well-lighted room. Each cone of flame should be steady and sharp. Adjust or clean burner if flame is yellow-orange.

FIGURE 3.1   Effective use of headings showing organization by user tasks.

Source: From *Use and Care Manual, Frigidaire Gas Range Model ES510 Control, Self-Cleaning Oven with Deep Well Style Cooktop*, Electrolux Home Products, Inc., Augusta, GA, p. 7. Used by permission.

Let's say that your company makes five models of snow blowers, ranging from a pint-sized, single-stage model suitable for clearing two or three inches of snow off a short sidewalk, to a heavy-duty, two-stage model that will chug right through the deepest drift and pitch the snow 50 feet. Chances are that some aspects of the five different models will be very similar, if not identical—how to turn the discharge chute to throw the snow in different directions, or how to mix gasoline and oil to achieve the proper fuel ratio, or safety precautions for use on uneven terrain or gravel driveways. Others will be different. If you write each section as a more-or-less self-contained module, then when it's time to develop a manual for a new model, you can easily piece much of it together from previously written modules.

This approach is made much easier by the ability to store text electronically. Electronic files take up little space and are easy to manipulate, making editing and revision much quicker. However, the ease of saving and editing creates its own problems—primarily of organization and record keeping. If you are going to assemble a manual from "parts" so to speak, you need to have a good parts catalog! Develop a system to keep track of what modules go with what products, when they were last revised, and so on. Without such a system, electronic document storage can easily get totally out of control and become useless. With a good system, the time—and therefore cost—required to develop a new manual when a model change takes place is significantly reduced. And modular construction of manuals is ideally suited to agile manufacturing, where every product is more or less a custom product.

DESIGNING FOR THE TWO-PAGE SPREAD

Related to modular design is the concept that you should design a manual as a series of two-page spreads. A two-page spread is all of a manual that a user can see at one time. Note that the key word is all they can *see* at one time—not all they can *read* at one time. Manuals (the good ones, anyway) are not made up of words alone. They include charts, photographs, illustrations, tables, and other primarily visual types of communication. In fact, some manuals are almost entirely made up of pictures. Seeing (rather than reading) is often a more accurate description of how people actually use manuals. Just as a cook making beef burgundy will frequently refer to the cookbook open on the counter while browning the meat and chopping onions, most people use manuals at the same time as they are doing something else: adjusting something, assembling something, operating something. That multitasking makes it even more critical to think in terms of two-page spreads because many times the user will have the manual lying open (perhaps weighted down with a wrench) while using his or her hands to perform some task on the product. Flipping pages is difficult when your hands are otherwise occupied.

What does designing for a two-page spread mean in practical terms? Obviously, some descriptions and procedures are complex enough that you cannot fit them all into two pages. Others may take up less than two pages. It simply means that you must keep in mind that regardless of actual length, from the reader's perspective, the material is being presented two pages at a time, no more, no less. The better you can chunk the material into two-page "bites," the better the design will fit reality. If you have a long procedure, see if you can divide it into more conveniently sized parts or phases. Try to ensure that diagrams or illustrations referenced in the text appear on the same or facing page, not overleaf.

MAKE THE LANGUAGE ACCESSIBLE

Many companies aim to write their manuals at a fifth-grade reading level to ensure that the majority of users will be able to understand the language. Reading level is certainly one element of what makes a manual easily understandable, but it is not the only one. Other elements that you need to consider are your users' experience base, their familiarity with the vocabulary of the industry, and your own use of product terms. Let's start with reading level.

LANGUAGE LEVEL AND READABILITY INDICES

If you pick up a paperback children's book and look at the back cover, you will often see a numerical score that indicates reading level in terms of school grade. For example, a reading level of 4 indicates fourth-grade level. How are these determined? It turns out that there are literally dozens of methods. Some, like the readability indices discussed in Chapter 2, rely solely on sentence length and syllable count. Here, for example, is a simplified version of the formula for the SMOG index ("Simplified Measure of Gobbledygook") developed by G. Harry McLaughlin:[11]

1. Select 30 sentences from a text, ten each from the beginning, the middle, and the end. Each group of ten should be made up of consecutive sentences.
2. Count all the words containing three or more syllables.
3. Take the square root, and add three.

The result gives you the approximate grade level of the material. Other measures include more complex indicators such as syntactic complexity of the sentences or how common the terms are. ("Together" and "manifest" are both three-syllable words, but the first is far more commonly used, and therefore more apt to be understood than the second.)

Readability calculators can be useful, but they should be a starting point, not the final arbiter of whether your manual uses an appropriate language level. Even more important than reading level is knowing your user.

WHO ARE YOUR USERS? ARE YOU SURE?

Chapter 2 addressed user analysis extensively, and there is no need to repeat that information here. But it is important to recognize that just because your user is familiar with the product or industry—the so-called "sophisticated user"—that does not necessarily mean that you can assume a higher level of language comprehension. Knowing whether your readers are sophisticated users is primarily useful for helping you decide what information you need to include, not for choosing an appropriate language level.

But what if your users are college-educated? What if your product is a mass spectrometer or other scientific instrument? You wouldn't want to insult your readers by pitching the manual at too low a reading level. Surely you would be safe writing the manual at higher than a fifth-grade level, wouldn't you? Why not write a sentence such as this:

> The introduction of solvent vapor or moisture into the analysis chamber may result in contamination with concomitant reductions of resolution and sensitivity and deteriorated signal-to-noise ratios, precluding accurate readings, thus emphasizing the importance of maintaining sample ions in a clean environment.

Even if your readers are capable of reading such a sentence, it takes work to decipher, and they probably will not want to read page after page of sentences like it. Your goal as a technical writer should be to remove as many barriers as possible between the information and your reader. Most people find it most comfortable to read at a level that is below the maximum they can manage.

It is possible to drop the reading level so far that it does become insulting—when your manual starts to sound like a first-grade reader, it's time to take a second look:

> Keep solvent vapor and moisture out of the analysis chamber. They can contaminate the sample environment. Contamination can reduce resolution. Contamination can reduce sensitivity. Contamination will cause poor signal-to-noise ratios. Contamination will cause inaccurate readings. Keep sample ions clean.

Too low a reading level can produce text that is just as exhausting to read as if it were too high. The good news is that there is a lot of space between those two extremes. For most readers, somewhere between fifth- and eighth-grade reading level is the best. Taking the same passage and putting it at about the right level would look like this:

> Keep solvent vapor and moisture out of the analysis chamber to avoid contaminating the sample environment. Contamination can reduce resolution and sensitivity and will cause poor signal-to-noise ratios. Because contamination will cause inaccurate readings, it's important to keep sample ions clean.

This version divides the information into three sentences, instead of cramming it all in one sentence (as in the first version) or chopping it up into six sentences (as in the second). This version uses the simpler language and verb-based active voice of the second version, but uses slightly more complicated sentence structure—allowing the syntax itself to convey the relationship of ideas. Doing so makes for a more comfortable read.

Choosing the right reading level is half the battle; choosing the right vocabulary is the second challenge.

JARGON, SLANG, AND TECHNICAL LANGUAGE

Every industry has jargon. Jargon is simply convenient shorthand for often-used technical terms or phrases. For example, in the fire service, the high-pressure "booster" hose that is carried on a reel on fire engines and brush trucks is often called the "red line," simply because the outer rubber covering on the hose is nearly always red. In real estate sales, houses that are put on the market without using a real estate agent are called "fizbos"—from the initials of "For Sale By Owner." Not all jargon, however, is understood throughout a particular industry. Sometimes there are regional differences. To use the fire service as an example again, a truck designed primarily to carry large quantities of water is called a "tanker" in the East and Midwest. In the Southwest, the same vehicle is called a "tender"—and if you ask for a tanker to respond to a fire scene, you'll get an airplane that carries water and/or fire retardant! Even more often, jargon is idiosyncratic to an individual workplace—and quite incomprehensible everywhere else. For this reason, you should avoid using jargon in a manual—even if your users are familiar with the product type, they may not know your company's jargon.

Equally out of place in a manual is slang, but for a different reason. Slang is defined as informal, idiomatic expressions that can be widely used or confined to specific groups, and that typically change over time. We're familiar with slang terms like "cool" (meaning "good") or "open 24/7" (meaning "always open"). Sometimes these slang expressions get absorbed into the language and over time become standard. Sometimes they die a quick death. What is true about slang and idioms in general, however, is that they usually are not precise and do not translate easily, especially if they are relatively new. Consider this sentence, taken from an actual manual: "Beef up the supporting wall." What exactly does that mean? Even if you know that "beef up" means to strengthen, it doesn't say exactly what to do. A better choice would be to specify the means to reinforce the supporting wall.

A third category of problematic terminology is technical language. Technical language may be entirely appropriate in a manual, depending on the audience. And it is usually able to be translated precisely. The only difficulty is making sure that your audience understands the terms you are using and that you are using them because you need the precision—not just to make it sound more impressive. Do you really need to call it a "threaded fastener?" Or would it work just as well to call it a "screw?" Your goal should be to write as simply and precisely as you can.

THE IMPORTANCE OF CONSISTENCY

English is a language with a lot of words, more than most other languages. In the world of manual writing, this rich vocabulary is both a strength and a weakness. It's a strength because it allows us to write very precise descriptions that are also very concise—we can find just the right word to say what we want. It's a weakness because we can use different words to describe the same thing—and that's confusing for users. For example, the same part might be referred to as a "spacer" in one location and a "shim" in another. We might refer to the same tool as a "hex key" on page 25 and an "Allen wrench" on page 30. Sometimes these inconsistencies are a result of more than one writer working on the same manual (a downside of the modular approach discussed earlier), and sometimes they slip in because a single writer uses both terms routinely. Whatever the cause, the result is confusion.

Readers tend to assume that writers use words intentionally, so if they see "spacer" in one place and "shim" in another, they will expect that these are different parts. Sometimes the words are similar—but not quite the same. For example, in a manual for a jackhammer attachment, the user is directed in one section to "tighten the air hose coupler" and in another section to "turn the air hose coupling." Are these the same part? Is "tightening" the same as "turning?" A first-time user would be unsure. And that unsure (and doubtless irritated) user would still have to decide what to do in response to the directions in order to use the product.

The difficulties are only compounded when the manual is translated, because now the decision rests with the translator. Are these parts one and the same? If so, the translator will probably use the same term. If not, the translator will use different terms, which are almost certainly not going to be as closely aligned as "coupler" and "coupling" are in English.

The more familiar you are with the product and how it works, the harder it will be for you to catch these inconsistencies. Whether you see "coupler" or "coupling" in the text, you will have a mental image of the correct part, because you know the product. It's always a good idea to have someone not familiar with the product read your draft. You'd be surprised how many inconsistencies they'll catch.

WRITING STRATEGIES THAT WORK

Once you have figured out what information to include, developed a plan for how to organize it, and settled on appropriate language and reading level, you still have to write sentences and paragraphs. If you're working as a technical writer, this might seem as though it should be the easiest part—after all, you're a writer. Yet it often seems to be a particularly daunting task. Remembering a few specific strategies can help make your writing easier and quicker. These strategies are organized into three groups:

- Techniques to divide and sequence information.
- Techniques to present information.
- Techniques to link information.

DIVIDING AND SEQUENCING INFORMATION

Information can be divided up in a variety of ways, just as different methods can be used to decide what drawer to use for which cooking utensils.* You can organize utensils by function: all the knives in one drawer, all the mixing spoons and whisks in another. You can organize by how often you use a particular utensil, so you end up with a drawer of tools you use every day, and a different drawer with those you use only occasionally (the turkey baster and the egg separator go in that one). You can even, as I did once, organize by color: utensils with red handles all go in one drawer, those with black handles in another.

For operator and service manuals, dividing information up functionally is logical and almost inevitable. This chapter discussed organizing macro-level information

---

* If all your cooking utensils fit in one drawer, the earlier cookbook analogy was probably lost on you!

by user questions—that is, functionally. Each user question asks how to do something, which is just another way of saying, "perform a function." Even smaller-level sections are likely to be organized functionally. For example, the manual for an all-in-one printer/copier/scanner might have several subsections within the overall category of "How to Use Your Printer" such as these:

- Loading paper
- Replacing ink cartridge
- Changing copier settings
- Scanning to a computer
- Clearing a paper jam

All of these answer a more specific "How do I . . .?" question.

One caveat is in order here. Be careful that when you organize information functionally, you do so (if possible) in terms of *how the user does something*, and not in terms of *what a product part does*. If you are writing a manual for a handheld marine GPS/two-way radio, avoid an organizational structure that groups items like this:

- Using the SQL Key
- Using the CLR Key
- Using the Soft Keys
- Using the Arrow Keys
- Using the Alpha-Numeric Keypad

While these appear to be functional divisions, they are not based on user tasks. Although the user might mentally ask, "What's this [left soft-key] for?" he or she doesn't really want a list of the operations that key will perform in various modes. Such a list is difficult to understand (no context) and even more difficult to remember. If you work with a design engineer as your subject-matter expert, you may find that the raw information you receive is organized by product part function. Remember, engineers want to tell you how the product works, not how to use it.

In some circumstances, a functional description of the product is appropriate. Particularly in a service manual, but sometimes in an operator manual, you may want to explain the operating principle of a machine or other product because the user needs to understand it in order to service it or (sometimes) use it. In those situations, describing what a particular part does is quite appropriate. Just make sure that you don't fall into the trap of only explaining how it works—and not how to use it.

Sequencing information is nothing more than deciding what to say first, what to say next, and what to say after that. Two strategies are most commonly used to sequence information:

- Spatial
- Chronological

Here's how each is used.

Spatial

Use spatial sequencing to describe a product in terms of its parts—particularly large products. For very large products, such as industrial machines, trucks, tractors, cranes, and even motorcycles, users say they approach the product with a preconceived spatial logic. For example, they may think of the product from front to back, from top to bottom, or from left to right when facing the product. Of course, variations of this basic perception will be required in specialized manuals or service instructions. In any case, if you use spatial orientation to sequence your information, be sure to visualize the product from the point of view of the product user (or service technician). For example, the logical view for service manuals on automobile exhaust systems would be to show and describe the system as seen from below by the mechanic working with the system overhead on a lift. For a printing press or a veneer dryer or a paper converter, a logical view would be to work from the infeed side to the output side of the machine.

Whatever the product or process, it helps to ask your user (or surrogate), "How do you think of this? Do you stand in front or at the side? Do you think of this from front to back? Top to bottom?" Then arrange the manual to match the user's spatial perception.

Chronological

Chronological organization is useful both for descriptions of processes and for instructions for performing procedures. Most processes and procedures have an inherent chronology; that is, the steps for doing something grow naturally out of the way the product or process works. For example, the user will usually want to know about setup or assembly before learning about operating procedures, maintenance, or storage.

However, at the level of paragraph or subsection, exceptions to strict chronology are quite common—and sometimes required. For example, suppose you are describing the operation and use of a home whirlpool. The dangers posed by high water temperatures to the elderly and to people with heart conditions or high blood pressure need to be mentioned, both in the manual and on the product, *before* the owner uses the whirlpool. Before you tell the user to remove the snap ring that holds a spring under compression, be sure to explain how to keep the spring from flying across the room. In short, if any step or procedure can, in its execution, cause injury to the user or damage to the product, be sure to explain this before the step is listed. Anticipate trouble spots in procedures, even if chronology is interrupted. Always warn of troubles or dangers before it is too late for the user to do something about them.

Rather than use these strategies in isolation, most manuals use a combination of writing strategies such as spatial or chronological sequencing. Figure 3.2 shows a manual that combines several strategies to convey sequential information.

Presenting Information

How you present information to the reader can make a big difference in how easy it is to follow and understand—even if every sentence is grammatically correct and every piece of information is sequenced perfectly. The reasons have to do with how our brains work and the way manuals are used.

UNPACKING AND CLEANING

Carefully unpack the saw, stand and all loose items from the carton. Remove the protective coating from the saw table surface. This coating may be removed with a soft cloth moistened with kerosene (do not use acetone, gasoline or lacquer thinner for this purpose). After cleaning, cover the table surface with a good quality paste wax.

ASSEMBLY INSTRUCTIONS

WARNING: FOR YOUR OWN SAFETY, DO NOT CONNECT THE SAW TO THE POWER SOURCE UNTIL THE SAW IS COMPLETELY ASSEMBLED AND YOU HAVE READ AND UNDERSTOOD THE ENTIRE OWNERS MANUAL.

ASSEMBLING STAND

1. Assemble the two top side braces (A) Fig. 4, which are 16-1/2" long, and the two top front and rear braces (B), which are 19" long, to the four legs (C) using the sixteen 5/8" long carriage bolts, flat washers and hex nuts supplied. **NOTE:** The top lips of the two top side braces (A) must fit on top of the top lips of the front and rear braces (B). The side braces (A) have holes on top for mounting the saw to the stand. Only tighten hex nuts finger-tight at this time.

2. Assemble the two bottom side braces (D) Fig. 4, which are 20" long, and the two front and rear braces (E), which are 22-1/2" long, to the four legs (C) using the sixteen 5/8" long carriage bolts, flat washers and hex nuts supplied. Only tighten hex nuts finger-tight at this time.

3. Assemble the four rubber feet (F) Fig. 4, to the bottom of each leg (C) as shown.

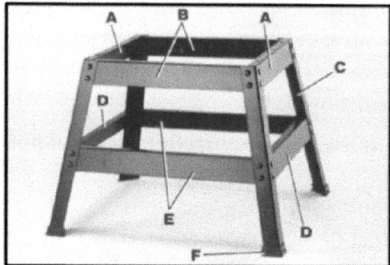

Fig. 4

ASSEMBLING SAW TO STAND

1. Position the saw (B) on the stand as shown in Fig. 5, lining up the four holes on the bottom of sides of the saw cabinet with the four holes in the two top side braces, one of which is shown at (A).

2. Fasten the saw to the stand using the four 5/8" long hex head screws, flat washers and hex nuts supplied.

3. After saw is assembled to stand, firmly tighten all stand mounting hardware.

Fig. 5

FIGURE 3.2 Combined writing strategies: general to specific, paragraph clusters, chronology. The sequence of pages follows the natural order of unpacking, cleaning, and assembly. Paragraph clusters are placed close to relevant visuals. Instructions use present-tense, active verbs. (*Continued*)

Source: From *10" Table Saw (Model 34–670) Instruction Manual*, Part No. 1340213, Delta International Machinery Corp., Pittsburgh, PA, 1995, pp. 5–7. Used by permission.

ASSEMBLING BLADE RAISING AND TILTING HANDWHEELS

1. Assemble the blade raising handwheel (A) Fig. 6, to the blade raising screw (B) making sure the slots (C) in the hub of the handwheel are engaged with the roll pins (D) on the raising screw shaft.

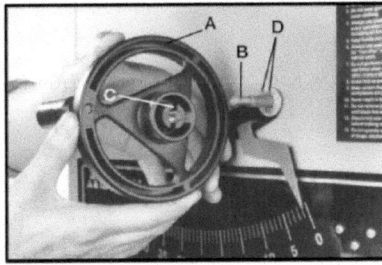

Fig. 6

2. Screw lock knob (E) Fig. 7, on end of raising screw shaft.

3. Assemble tilting screw handwheel (F) and lock knob (G) Fig. 7, to the blade tilting screw shaft in the same manner, as shown in Fig. 7.

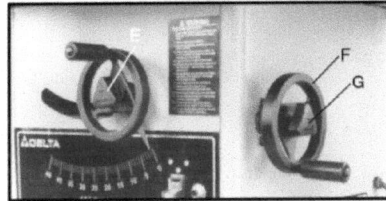

Fig. 7

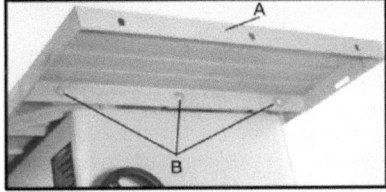

ASSEMBLING EXTENSION WINGS

1. Assemble extension wing (A) Fig. 8, to the saw table using the three screws and washers (B). With a straight edge (C) Fig. 9, make sure the extension wing is level with the saw table before tightening the three screws (B) Fig. 8.

2. Assemble the other extension wing to the opposite end of the table in the same manner.

Fig. 8

Fig. 9

FIGURE 3.2 Combined writing strategies: general to specific, paragraph clusters, chronology. The sequence of pages follows the natural order of unpacking, cleaning, and assembly. Paragraph clusters are placed close to relevant visuals. Instructions use present-tense, active verbs. (*Continued*)

Source: From *10" Table Saw (Model 34–670) Instruction Manual*, Part No. 1340213, Delta International Machinery Corp., Pittsburgh, PA, 1995, pp. 5–7. Used by permission.

ASSEMBLING SAW BLADE

1. Make certain the saw is disconnected from the power source.

2. Remove the table insert (A) Fig. 10.

3. Raise the saw blade arbor (B) Fig. 10, to its maximum height by turning the blade raising handwheel counter-clockwise and remove the arbor nut (E) and flange (D) from the saw arbor.

4. Assemble the saw blade (C) to the saw arbor making sure the teeth of the blade point down at the front of the table, as shown in Fig. 10, and assemble the flange (D) and arbor nut (E) to the saw arbor and tighten arbor nut (E) as far as possible by hand, being sure that the saw blade is against the inner blade flange.

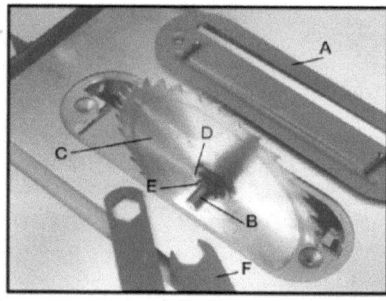

Fig. 10

5. Using the open end wrench (F) Fig. 10 and Fig. 11, supplied, place the wrench (F) on the flats on the saw arbor to keep the arbor from turning and tighten arbor nut (E) using the remaining wrench (G) Fig. 11, by turning the nut counterclockwise.

6. Replace table insert (A) Fig. 11, making certain that it is flush with table surface.

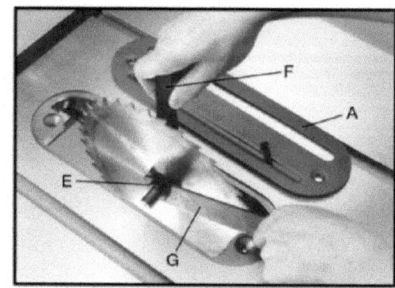

Fig. 11

ASSEMBLING GUIDE RAILS

1. The guide rail (A) Fig. 12, with graduations is to be assembled to the front of the saw table with the gradua-tions up.

2. Insert the two special screws (B) Fig. 12, through the two holes (C) in the guide rail and place spacers (D) between the guide rail (A) and saw table. Thread the two special screws (B) into the tapped holes in the saw table. Do not completely tighten the two screws (B) at this time.

3. Insert special screw (E) Fig. 12, through hole (F) in guide rail and place spacer (G) between guide rail and extension wing. Fasten with flat washer, lock washer and nut (H). Tighten three screws (B) and (E) to fasten guide rail to table and extension wing.

4. Assemble the remaining guide rail to the rear of the table in the same manner.

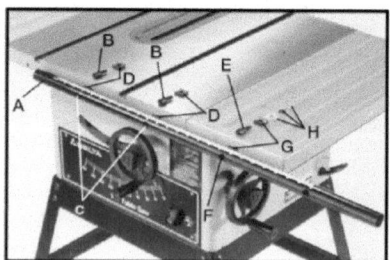

Fig. 12

FIGURE 3.2   (Continued) Combined writing strategies: general to specific, paragraph clusters, chronology. The sequence of pages follows the natural order of unpacking, cleaning, and assembly. Paragraph clusters are placed close to relevant visuals. Instructions use present-tense, active verbs.

Source: From *10" Table Saw (Model 34–670) Instruction Manual*, Part No. 1340213, Delta International Machinery Corp., Pittsburgh, PA, 1995, pp. 5–7. Used by permission.

Our brains are designed to see patterns and to prefer structure. That's why, for example, it's easier to remember a telephone number, complete with area code, than to remember a random 10-digit number. It's easier to remember three separate pieces of information if we can see them as parts of a whole. This human tendency to look for and see patterns makes good survival sense: if we discover a pattern, we can predict what comes next. As designers of manuals, we can take advantage of the brain's desire for structure and pattern to make information easier to absorb.

Manuals are almost never read in isolation. People almost always read manuals while they are also interacting with the product. It's common for manual writers to instruct their readers to "read through all the instructions first, before beginning the installation." No one does that. Research cited by Ganier shows that "procedural documents are rarely ever used in this manner."[12] What actually happens, as Ganier notes, is that users are engaged in a very complicated activity involving the instructions, the product, the user's prior knowledge, and other user characteristics, such as ability and intelligence. The result is that the user tries to juggle all these factors in working memory—and as a result, tends to read a single instruction, do the directed action, then read the next, and so on. Working memory just doesn't have the capacity to manage much more.

At the same time, our brains work to construct a mental model of what the goal is of what we're doing. Have you ever found yourself dutifully following instructions for assembling a piece of furniture or a propane grill without knowing why the instructions had you performing certain actions? It's very uncomfortable. When you finally got far enough along to be able to say, "Oh, I get it," you probably felt much more relaxed and confident. All along, you had been trying (unsuccessfully) to create a mental model of where you were going. As Ganier notes, the "accuracy, effectiveness, and speed of elaboration of the mental model" depends in part on how the instructions are presented.[13]

Most manuals use a variety of ways to present information, including verbal, visual, and sometimes even interactive forms. Chapter 5 and Chapter 6 address visuals in manuals and non-traditional learning tools. Even using words alone, however, certain strategies will make it easier for your users to absorb information. Three particularly useful techniques are

- Big picture first
- Parallel patterns
- Power of the list

Each of these techniques helps the user to formulate the mental model and minimize the strain on working memory.

Big Picture First

Manuals are not novels—and they are especially not mystery novels. In a good "whodunit," you don't find out for sure who the villain is until the last few pages. The author and the reader play a little game: the author lays out clues, some of which are intended to mislead, and the reader keeps forming different mental models to accommodate the new information. The writer's goal is to fool the reader until the end, when all is explained. Contrast that with how a newspaper article is put together. It's just the opposite: while the mystery novel waits to the last page to reveal that the butler bludgeoned Lord Pinkerton with his own lead-filled shooting cane, the newspaper leads with a three-inch headline screaming "BUTLER GUILTY!" Every essential fact will appear in the first paragraph of the newspaper article—the proverbial who, what, when, where, and why. As you read further, more and more detail is added, but no entirely new information.

The difference in strategy reflects the difference in purpose: mystery novels are intended to entertain and provide a relaxing way to spend time, whereas newspapers are intended to inform. It would be counterproductive for a newspaper to scatter details at the beginning of the story and only bring them all together at the end. The readers—seeking knowledge rather than diversion—would be frustrated and confused until they finally got through the whole article. Most of them probably wouldn't bother to read that far.

Manuals are like newspapers in that the aim is to give users needed information. It makes sense, then, that manuals should start with the big picture and add details as the reader gets further into each section. This general-to-specific principle works for the manual as a whole and it also works for individual chapters, subsections, and even paragraphs. Figure 3.3 illustrates this principle in action.

Notice how the entire manual begins with a big-picture look at the product ("Printer Overview"), and the most detailed information ("Specifications") is left to the end. Even within sections, the general-to-specific order is maintained. The section called "Setting Up the Printer," for example, begins with an introduction and proceeds with instructions for five tasks, some of which are divided into smaller subsections. If you are reading the instructions for how to "load media for peel configuration," even if you don't know exactly what the title means, you know that this procedure is part of "Loading Media," which itself is a part of "Setting Up the Printer."

Another way to think of it is to compare organizing a manual to a digital map with an interactive zoom. At the farthest-out zoom level, you may be looking at an entire state with a few major cities and highways visible. As you zoom in, you are given more and more detail, until at the farthest-in zoom level you can see individual street names and maybe even house numbers. That extreme level of detail can be very helpful—but if that's all you can see, it would be difficult to navigate from, say, Minneapolis to Chicago. Users of manuals need to be able to see the big picture before they can make sense of the details.

Parallel Patterns

As noted, the human brain is predisposed to look for patterns. Good writers take advantage of this preference for patterns by using similar grammatical structures to convey similar ideas. This technique is called using "parallel structure." When

Contents

Printer Overview

Getting Started

Setting Up the Printer

Using the Front Panel

FIGURE 3.3 Table of contents showing general-to-specific structure. (*Continued*)

Source: From *E-Class Operator's Manual*, Datamax Corp., Orlando, FL, pp. i–iii. Used by permission.

FIGURE 3.3 (Continued) Table of contents showing general-to-specific structure. (*Continued*) *Source:* From *E-Class Operator's Manual*, Datamax Corp., Orlando, FL, pp. i-iii. Used by permission.

FIGURE 3.3 (Continued) Table of contents showing general-to-specific structure.

Source: From *E-Class Operator's Manual,* Datamax Corp., Orlando, FL, pp. i-iii. Used by permission.

procedures or mechanisms are closely related, parallel sentence strategies also serve to sharpen focus. Notice the difference in clarity in the following examples.

Example 3.2: Non-Parallel vs. Parallel Instructions

Non-Parallel

> Install front bolts with the threads down.
> On the rear bolts, make sure the threads face up.

Parallel

> Install front bolts with threads down.
> Install rear bolts with threads up.

Readers can understand the first, but they will grasp the second more quickly. If you are trying to make comparisons or contrasts between steps or characteristics of a process or product, you can also use parallel structure to heighten the comparison.

Example 3.3: Non-Parallel vs. Parallel Comparisons

Non-parallel Comparison

> More modern devices have electronic controls, whereas they were formerly operated mechanically.

Parallel Comparison

> Modern devices are electronically controlled, whereas formerly they were mechanically controlled.

The non-parallel example is comprehensible, but the comparison is not sharp. Using parallel structure makes the comparison much easier to see.

Parallel structure is essential when you are presenting a series of steps in a procedure or parts in a sequence. Using parallel structure allows the reader to concentrate on understanding the content of the instructions rather than figuring out the sentence structure. If all the instructions follow the same pattern, the reader only has to puzzle it out once. Example 3.4 illustrates the importance of consistent verb forms.

Example 3.4: Inconsistent vs. Consistent Verb Forms

Inconsistent Verb Forms

> Always wait for the tractor to come to a complete stop. After lowering the mowing deck to the ground, make sure the transmission is shifted to the N position; the park brake should be set to prevent the tractor from rolling. Then remove the key.

Consistent Verb Forms

> Always wait for the tractor to come to a complete stop, then lower mowing deck to the ground, shift the transmission to N position, set the park brake so the tractor will not roll, and remove the key.

Readers will understand the first version, but they will have to slog through shifts from active to passive voice and unnecessary glitches in verb tense sequence ("wait," "after lowering," "is shifted," "should be set," "remove"). In the second version, each element of the series begins with an imperative verb ("wait," "lower," "shift," set," "remove"). The series of commands presents a consistent, predictable, comfortable pattern—in short, a clear and reassuring map of what to do.

Using a verb-first command pattern for instructions has a second benefit: it reduces ambiguity. If the manual says, "The humidistat can be calibrated by turning the screw located in the base of the unit," the reader is left with questions: Am *I* supposed to calibrate it? Or is this just additional information? By contrast, if the manual says, "To calibrate the humidistat, turn the screw located in the base of the unit," there's no doubt—that instruction is directed at the user.

Regardless of the context, anytime you can use sentence structure to establish a useful pattern for a reader, do so. You'll leave the reader free to focus on the content.

The Power of the List

A parallel pattern is even clearer—and easier for the reader—when the matching sentences are "stacked" in a list. The list strategy makes use of a powerful communication tool, the vertical arrangement. To illustrate the power of the vertical, add the following set of numbers (without a calculator!): 456 + 1678 + 45 + 789 + 9. Keep track of how long it takes you to do this.

```
Now add
 357
4789
  23
 540
   8
```

The number of units, tens, hundreds, and thousands is the same in both sets, but most people are far quicker with the vertical arrangement, because the tens, hundreds, etc., are visually next to each other. (As Chapter 5 discusses, the same principle determines whether to align items in rows or columns.) The list is the verbal equivalent of a column of numbers. In Example 3.5, the words "before" and "when" serve as predictable organizers, like units and tens, and the reader's eye picks up only the key words as it sweeps down the passage. Note also that the white space around each element makes the list even clearer.

Example 3.5: Linear vs. List Mode

Linear Mode

The system must be vented under the following circumstances: Before starting an engine that has not been operated for an extended period of time. When the fuel filters have been replaced. When an engine runs out of fuel in operation. When any connections between the pump and fuel tank have been loosened or broken for any reason.

List Mode

The system must be vented under the following circumstances:

- Before starting an engine that has not been operated for an extended period of time

- When fuel filters have been replaced
- When an engine runs out of fuel in operation
- When any connections between injection pump and fuel tank have been loosened or broken

Should you use numbers or bullets for a list? The answer depends on whether there is a natural order for the elements in the list or whether it doesn't matter what order they appear in. In Example 3.5, bullets are appropriate because it doesn't make any difference how you order the elements: any one of these conditions could occur with or without the others. By contrast, instructions for performing a procedure should always be presented as a numbered list, because the order in which you perform the operations does matter.

An additional advantage of using a list format—and a numbered list in particular—for presenting instructions is that it makes it easier for users to keep their place in the instructions. Remember that a person performing a procedure is almost certainly going to be looking back and forth between instructions and product. Putting instructions in a linear mode will make it far too easy to miss one as the eye jumps back and forth. Using a numbered list helps ensure that the user will follow all the steps in the proper order.

Presenting information in list form is not the total answer. You still have to follow the other principles of good communication presented here. Remember that readers need pattern and progress. An undifferentiated list of 45 steps in a procedure is not much better than a page-long paragraph of wall-to-wall text. That same list of 45 steps would be much easier to use if it were divided into five segments, each with a heading that signals the purpose of that subset of instructions. That way, the information is divided as well as sequenced, the headings give the reader the "big picture" of what the ultimate goal is and how each smaller section fits in, and the reader will have a much easier time finding his or her place on the page. The same principles hold true for other kinds of information that may be presented in lists. For example, it is common to have a "safety page" at the front of a manual with lots of safety warnings, often presented as bulleted lists. People will be much more likely to read them if they are grouped under logical headings:

- Workplace Safety
- Electrical Hazards
- Moving Parts Hazards

Just as modular organization can make it easier for a writer to assemble a manual, logical chunking of information can make it easier for a reader to absorb the content.

LINKING INFORMATION

Have you ever read a paragraph over and over, simply because you couldn't make it stick in your mind? You might know the meaning of each word and understand each sentence individually, but somehow you just couldn't make them fit together into a coherent package. Sometimes when that happens, the content is not logically organized within the paragraph, but more often it's caused by one of two writing problems:

- The focus is not consistent from one sentence to the next.
- The transition from one sentence to the next is not smooth.

Either of those problems will make it difficult for a reader to glue the content together—however well it's organized—into a comprehensible whole. As my writing students used to say, "It just doesn't flow." Fortunately, a few simple techniques can fix those problems. Like good product design, you don't notice them when they're in place, but you sure notice their absence!

Sharpening the Focus

If you've ever studied a language other than English, you have probably encountered situations in which word order differs from what is common in English. For example, in French and Spanish, often the adjective follows the noun it describes rather than precedes it. The English phrase "a red car" becomes *el carro rojo* in Spanish and *un auto rouge* in French. Many languages also assign gender to nouns. In the previous example, both *carro* and *auto* are "masculine" nouns. (Interestingly, a French synonym, *une voiture*, is feminine.) Some languages, such as German and Latin, also give nouns a different form depending on what role they play in a sentence. English does this only with personal pronouns: the difference between "she" and "her," for instance, is solely whether the word is being used as a subject or an object in a particular sentence. Here's an example: "She likes animals" versus "Animals like her." In the first sentence, the pronoun is the subject of the sentence; in the second, it's the object of the verb.

So what? If you have different forms for nouns depending on the role they play, then you can put the words in practically any order in a sentence and the reader can still figure it out. If you've ever struggled through a long sentence in German, you've experienced it first-hand. English, by contrast, is not so flexible. Unless a sentence uses personal pronouns, the meaning is highly dependent on word order. There's a big difference between "the cat chased the mouse" and "the mouse chased the cat"!

In English, the preferred word order for most sentences is subject—verb—object. This order feels "natural" to an English-speaker. Inverted word order (her, animals like) can be deciphered, but it's more difficult. Readers tend to look for the subject of a sentence to signal the content. If that focus changes from one sentence to the next, a paragraph can become very confusing.

Compare the passages in Example 3.6 to see how this works.

Example 3.6: Changing the Subject Changes the Focus

Focus Consistent

Manual reel-type push mowers are often a good alternative to gasoline-powered rotary mowers. Modern ones are much lighter than older models and are easy for almost anyone to use. Push mowers are quiet and easy to maintain. They do not burn gasoline and they do not pollute. These mowers cut grass blades cleanly, causing less stress on the plant, and deposit the clippings back into the lawn, where they can decompose and replenish the soil.

Focus Inconsistent

Manual reel-type push mowers are often a good alternative to gasoline-powered rotary mowers. Almost anyone can use a modern push mower easily, because of their light weight compared to older models. Maintaining a push mower is easy and they're quiet to use. You don't have to worry about the price of gasoline or polluting the air with fumes. Best of all, the grass plants will be less stressed because the blades cut cleanly and the clippings are deposited back to decompose and replenish the soil.

Most readers would find the first of these paragraphs much easier to read than the second, although the content and the sequence of information are the same in both. The first paragraph seems to "flow," while the second seems disjointed and confusing. What's the difference between them? In the first passage, the subject of every sentence is essentially the same: some equivalent of "push mower":

1. Mowers
2. Ones (referring to mowers)
3. Mowers
4. They (referring to mowers)
5. Mowers

In the second passage, every sentence has a different subject:

1. Mowers
2. Anyone
3. Maintaining
4. You
5. Plants

Most readers would not be able to put their finger on exactly what it was that made the first one easier to read than the second, but they would feel the difference, even in such a short selection.*

Not every paragraph lends itself to using the same subject for each sentence (this one, for example). Especially when you are describing a process or a mechanism, you would expect to shift the focus as you moved from one part to the next. The lesson is simply that the subject of the sentence dictates where the reader's focus will be—make sure your sentence structure points the reader in the right direction.

Smoothing the Transition

Transitions link one sentence to the next. They are the glue that helps make a paragraph coherent. Sometimes transitions arise naturally out of the way the content is connected together; other times the writer needs to help the reader move from one sentence to the next. Typical transition devices include

---

* The passage in Example 3.6 might work even better as a bulleted list.

- Repeated concepts
- Time or space cues
- Transitional words and phrases

Repeating concepts is a subtle but extremely effective way of linking sentences. The idea is simple: whatever word or concept appears at the end of one sentence appears right away at the beginning of the next. Like links in a chain, the repetition of ideas binds the two sentences together. See Example 3.7. Repeating concepts is a natural fit whenever you are describing any cause-and-effect sequence, such as a process or the function of a mechanism, since the effect of Event A in turn becomes the cause of Event B.

Example 3.7: Repeating Ideas to Link Sentences

- The Model 24 Hi-Flow pump is fitted with an intake screen. This screen prevents debris from entering the intake and damaging the impeller. [*screen—screen*]
- Tighten the adjusting nut with a 7/16″ box-end wrench. The wrench must be short enough to clear the frame. [*wrench—wrench*]
- This furnace is equipped with electronic ignition. If the igniter malfunctions, the system will lock out, shutting down gas flow to the burner. [*ignition—igniter*]

Time or space cues are more obvious to the reader than repeated concepts, but they can work well to link sentences together by helping the reader to understand the relationship between them. Time cues are words or phrases such as "before," "at the same time," "next," "then," etc. You may find that instead of using time cues to link the sentences in a linear paragraph, a better choice is to use a numbered list of steps to convey the same information. The numbering takes the place of time cues to show how the information relates. Space cues are words or phrases such as "under," "over," "behind," "inside," and so on, normally used to show the relationship of parts—and the sentences about them—to one another. As with temporal relationships, a linear paragraph may not be the best choice to convey spatial relationships. You may find that a well-designed illustration or photograph will do the work more effectively (and will not need to be translated). Chapter 5 discusses the use of graphics in manuals.

Transitional words or phrases such as "similarly," "by contrast," "for example," and "however" tell the reader overtly how ideas relate to one another. If you see the word "however" at the beginning of a sentence, you know that what follows will contradict or limit what was stated in the previous sentence.

Skilled use of transitions, along with solid organization and a consistent focus, help readers move smoothly from one sentence to the next. As you read through your draft (or better yet, have someone else read it), look for spots where the writing seems awkward or disjointed. It may be that you need to link the sentences together better. Usually when you are writing steadily and uninterrupted, transitions come automatically. When your writing is interrupted by frequent phone calls or other disruptions, and you have to write in bits and pieces instead of pages at a time, it's easy to lose your train of thought, and with it the easy flow from one sentence to the next. Find the rough spots and use transition techniques to smooth them out.

SUMMARY

Every writer has two fundamental tasks: deciding what to include and deciding what order to put it in. For manuals, the best guide for deciding what to include is to ask what the reader needs to know. Users come to products with questions, and a manual should answer those user questions—and answer them from the user's point of view. Because those user questions are sometimes very specific, it may be necessary to repeat information in more than one part of the manual—especially safety information, since no one is likely to read the manual from cover to cover, as if it were a novel. Manuals are more like cookbooks, in which information is arranged in self-contained chunks for easy access. Writing the manual as a series of stand-alone modules, rather like a cookbook, allows easy transition from one model of the product to the next, and designing for the two-page spread that the reader actually sees makes for efficient and easy use.

However carefully organized the manual is, it still must read easily as well. That means, at the least, that the manual must be written in language the users can understand. Too high a language level can put up enough of a barrier that people simply won't bother. Similarly, too much technical language or jargon can also put up a barrier to easy comprehension. Jargon is a particular problem in manuals that need to be translated, but equally problematic is inconsistent use of terms. Writers must be diligent in making sure that their manuals are written in readable, straightforward English. In the long run, that will ensure that whatever language they are translated into will also be readable and easy to understand.

Good organization and proper language level will go a long way to making manuals usable, particularly when combined with effective writing strategies such as logical paragraph organizers, presenting information so that the reader gets the big picture before being inundated with details, or using numbered lists for presenting instructions. Equally important is making sure that the writing "flows" with a consistent focus and smooth transitions from one topic to the next.

The bottom line is that users decide very quickly whether reading a manual is worth the trouble. If it does not seem worth it, they'll find a way around it. As a writer, you need to make using the manual the easy choice. The organizational and writing strategies presented in this chapter are part of the answer to that problem. The rest lies in designing the manual so that it is inviting to read and easy to navigate. The next chapter goes into more detail on how headings and other structural cues can help the reader map information in a manual. Those techniques combine with the basic principles presented here to achieve the goal of giving users the information they need to use the product—and use it safely.

CHECKLIST: THE USER-FRIENDLY MANUAL

The following is a list of questions to ask yourself about the writing strategies for your manual:

- Have I identified what's important—from the user's point of view?
- Have I focused on need-to-know information?

- Have I repeated important information?
- Have I given cues, such as headings, overviews, summaries, and transitions, to help the reader find his or her way through the manual?
- Have I made it easy for the user to skip around in the manual?
- Is my manual more like a cookbook and less like a novel?
- Have I presented sections in easy-to-use modules? Two-page spreads?
- Have I ordered the sections in a logical sequence?
- Have I eliminated jargon, slang, and unnecessary technical language?
- Will the manual be easy to revise and/or translate?
- Have I warned of safety issues before a hazardous step?
- Have I answered the user's questions?

NOTES

1. Ruth V. Small, Bernard J. Dodge, and Xiqiang Jiang, "Dimensions of interest and boredom in instructional situations." Proceedings of Selected Research and Development Presentations at the 1996 National Convention of the Association for Educational Communications and Technology, Indianapolis, IN, 1996, pp. 712–726.
2. Alan Cooper, *The Inmates Are Running the Asylum: Why High-Tech Products Drive Us Crazy and How to Restore the Sanity* (Indianapolis, IN: SAMS, a division of Macmillan Computer Publishing, 1999), p. 156.
3. Karen Schriver, *Dynamics in Document Design* (New York: John Wiley and Sons, Inc., 1997), p. 222.
4. Cooper, *The Inmates Are Running the Asylum*, p. 11.
5. Donald A. Norman, *The Design of Everyday Things* (New York: Doubleday/Currency, 1990), originally published as *The Psychology of Everyday Things* (New York: Basic Books, 1988), p. 173.
6. Cooper, *The Inmates Are Running the Asylum*, pp. 19–20.
7. Shriver, p. 222.
8. Franck Ganier, "Factors Affecting the Processing of Procedural Instructions: Implications for Document Design," *IEEE Transactions on Professional Communication*, 47, no. 1 (March 2004): 15–26, p. 15.
9. Stacy Schiff, quoted in Thomas Washington, "We're on Information Overload," *The Christian Science Monitor*, February 6, 2008, p. 9.
10. Donald C. Woolston, Assistant Dean (Ret.), Engineering General Resources, University of Wisconsin-Madison.
11. G. Harry McLaughlin, "SMOG Grading: A New Readability Formula," *Journal of Reading* (May 1969): 639–646.
12. Franck Ganier, "Factors Affecting the Processing of Procedural Instructions: Implications for Document Design," *IEEE Transactions on Professional Communication* 47, no. 1 (March 2004): 15–26.
13. Ganier, "Factors Affecting Procedural Instructions," 17.

4 What You See Is What You Read. . . . Or Not

OVERVIEW

The first edition of this book back in 1984 was titled *Writing and Designing Manuals* because at that time those were two separate functions performed by different people. Back then, the norm was that writers wrote text, illustrators drew art, and designers—sometimes called "pasteup artists"—put the two together into page layouts. Indeed, the designers often literally "pasted up" page layouts, cutting "camera-ready" printed text and art with a razor blade or an X-Acto® knife and fitting it together on a "mechanical," which then was photographically reproduced for printing. Photographs, or "continuous-tone" art, had to be shot through a screen that converted the image to black dots of various sizes and spacings. The viewer's eye then would merge the dots and see them as shades of gray, the same way our eyes blend dots of color in an Impressionist painting. A published manual represented the combined efforts of a crew of specialists.

The computer changed all that. In less than 25 years, an entire industry was transformed, and terms like "halftone screen" and "Linotype operator" now seem as quaint as buggy whips. Now writers can choose a font with the click of a mouse button, artists can resize or reorient images electronically, and photos can be scanned, digitized, and stored as electronic files. The writer, as likely as not, will be the one to put it all together in a page design—digitally, of course. What used to be a team effort of a half a dozen or more specialists may now be a collaboration between just two: a writer and an artist. (In some small operations, writer and artist may even be one and the same!) This change has provided both opportunity and challenge.

When the person who writes the chapter also chooses the typeface and lays out the pages, that person can ensure that the arrangement of the text and art on the page reflect the meaning. The writer understands the organizational framework of the text and knows where the illustrations need to be placed for the reader to make best use of them. Good page design gives the reader subliminal clues to the meaning carried in the words and pictures: the writer, who knows the meaning backwards and forwards, is an ideal choice to arrange those subliminal clues.

By now it should be clear that the designer of a manual has something of an uphill climb to produce a manual that will entice a user to read it. Couple a nearly universal disinclination to read manuals with many competing demands for attention and the result is a challenging problem: how can we design a manual to be sufficiently inviting that users will choose to read it, despite the obstacles in their way?

MAKE THE MANUAL LOOK EASY TO READ

Two of the reasons often given for avoiding manuals are that they're hard to read and they're confusing. These have as much to do with presentation as with substance. Just as a beautiful plate presentation can make an ordinary meal look more appetizing, good document design can make the content of a manual seem more palatable. While organized content and good writing are critical to making a manual useful to the reader, presentation is equally critical—if the manual's appearance turns the user away, the quality of the organization and writing will make no difference, for the user will never experience them.

Principles of Page Design

Like anything else, much of the art of good page design must be learned by experience. You try something, and it works—or it doesn't. You notice a particularly effective layout in a document and spend a moment to figure out what makes it so effective. Gradually, you begin to get a feel for how various design elements work together and how to translate that information into placement of the text and illustrations in your own manual. Nevertheless, just as with any other design process, certain fundamental rules apply. The most important of these is that form follows function. In other words, make your design choices on the basis of whether they make the manual work better. Remember that in designing a manual, you are creating a document with a very specific purpose: to convey to the user the information needed to properly and safely use the product. Everything else comes second. If a design element does not contribute to that fundamental purpose, eliminate it—even if it is artistic, eye-catching, and innovative.

If the primary rule is that form follows function, then it stands to reason that the simplest form that will do the job is the best. Remember that the purpose of format and layout is to help the reader understand the information by signaling its structure. If the design becomes too complex, it will obscure the very thing it is intended to reveal. Too complicated a format makes it harder for the reader to keep track of the various elements, and the reader will usually just give up trying. Most design innovations in products fall into two categories: those that improve performance and those that achieve the same performance with a simpler design. As long as your page design does its job, the simpler the better.

Product design engineers add parts to a product for specific purposes. Every nut and bolt, every spring, every lug, every piece of complexity is there to perform a specific function—if it is not needed, it is eliminated. The reason for this, of course, is that unneeded parts add to the cost of manufacturing and maintaining the product. In page design, "extra parts," that is, design elements that serve no function (or merely a decorative one), also add to the cost of the product, although in a more indirect way. Technical writing should be transparent. In other words, the writing (and by extension, the page design) should convey the content as directly as possible to the reader, without calling attention to itself. In technical writing, the reader generally should not even be consciously aware of the language used or the design adopted. The reader's focus should be on the information being conveyed, rather than the form in which it is packaged.

If that is not the case, and attention is drawn to sentence structure or format, the reader is less likely to acquire the needed information to use the product properly and safely. Costs appear in the form of calls to the 800 number for technical support or, worst case, a products liability lawsuit. Is it a bit of a stretch to imagine that a flawed page design could be the ultimate cause of a lawsuit? Yes and no. It is unlikely that page design alone would be at fault. However, it could easily be included as one more problem area in a faulty manual, rendering the product "defective."

Clearly, if the purpose of a format is to provide the reader with clues to how the material is organized, the format needs to be consistent or it won't work. If the writer uses 14-point bold headings in one place and 12-point italics in another to signal the same level of hierarchy in the organizational structure, the reader will be confused. Since most manuals are team efforts, it is important to agree on formatting decisions early on, so that whatever is chosen can be applied by all the writers working on a single document. Likewise, if the same format is to be used in several different manuals, you must make sure that all the writers apply the format the same way.

MAKING A GOOD FIRST IMPRESSION

Users decide whether to read a manual in a remarkably short time. You should assume your potential readers will make their decision whether to read your manual in less than three seconds. (Actually, it may even be much less than that.) What makes the difference between a manual that looks inviting and one that looks forbidding? Let's try an experiment. Figure 4.1 shows the first page from two different versions of the same manual. Look at the pages while counting "one thousand one, one thousand two, one thousand three," and then answer these questions:

1. Does one seem "friendlier?"
2. Does one seem more "businesslike?"
3. Does one look harder to read?
4. Would you prefer to read one over the other?

If you're like most people, you think Version 1 looks more "businesslike" and also more difficult to read, and you prefer Version 2 with its "friendlier" look. Not everyone will agree, but most people prefer the look of Version 2. How long did it take to make your assessment? In all likelihood, you didn't need a full three seconds. Potential readers assess the readability of documents very quickly, and usually unconsciously. They know what they conclude about a document, but they usually do not stop to analyze why they drew that conclusion.

Let's take a closer look at Figure 4.1 and spend a few moments analyzing the differences that would have been noticeable in three seconds or less. There are some differences in language level and sentence structure between the two, but those probably would not have been apparent in such a short time. What we can see—almost

DRI EAZ
Owner's Manual – DrizAir Dehumidifiers
DrizAir 80, DrizAir 1200, DrizAir 2000 LG, DrizAir 2400 LG
115-volt models

DRI-EAZ PRODUCTS, INC.
15180 Josh Wilson Road, Burlington, WA USA 98233
Fax: (360) 757-7950   Phone: (360) 757-7776   www.dri-eaz.com

DrizAir Dehumidifiers are commercial refrigerant dehumidifiers, designed for drying building structures and contents. They are designed for high performance water removal, as well as durability and ease of use.

Read And Save These Instructions

 ## Safety Information

Keep Children Away: Do not allow children to play with or around the unit, which could result in injury. Be sure the unit is inaccessible to children when not attended.

Keep Unit Grounded: Always operate the unit with a grounding plug and a grounded electrical outlet. A grounding plug is an essential safety feature that helps reduce the risk of shock or fire.

Protect Power Cord from Damage: Never operate a unit with a damaged power cord, as this may lead to electrical or fire hazards. If the power supply cord is damaged, it must be replaced by a cord of the same type and amperage rating.

Extension Cords: Extension cords must be grounded and able to deliver the appropriate voltage to the unit.

Handle With Care: Do not drop or throw the unit. Rough treatment can damage the components or wiring and create a hazardous condition.

Run on Stable Surface: Always operate the unit on a stable, level surface, like the floor or a strong counter, so it cannot fall and cause injury.

Secure During Transport: When transporting in a vehicle, secure the unit to prevent sliding and possible injury to vehicle occupants.

Keep Out of Water: Never operate the unit in pooled or standing water, as this may create a risk of injury from electrical shock. Do not store or operate outdoors. If electrical wiring or components become wet, thoroughly dry them before using the unit.

Keep Air Intakes Clear: Do not clog or block air intakes, as may occur if operated too close to draperies or similar materials. This may cause the unit to overheat and result in a fire or electrical hazard.

Keep Filter Clean: Always use a clean air filter. Do not allow any material to clog the filter, as this may cause the dehumidifier to overheat. Do not allow oil, grease, or other contaminants to be drawn into the dehumidifier.

Keep Electrical Components Dry: Never allow water inside the dehumidifier's electrical components. If these areas become wet for any reason, thoroughly dry them before using the dehumidifier.

Allow Repair Only by Qualified Person: Do not attempt to disassemble or repair the unit if you are not qualified to do so. You may handle some maintenance and troubleshooting, but make sure that more complex problems are handled by an authorized

service technician. For information about authorized repair, call Dri-Eaz at (360) 757-7776.

HOW TO USE DEHUMIDIFIERS

DrizAir Dehumidifiers are designed to reduce water vapor or high humidity in a building or part of a building. The purpose is to prevent humidity damage, and to dry out wet materials such as carpet, carpet cushion, floors, walls, furniture, contents, lumber, and structural materials.

To speed up the rate of evaporation, and dry out materials faster, use Dri-Eaz TurboDryer airmovers. The DrizAir Dehumidifier will be needed to remove the additional water evaporated by the TurboDryers.

Positioning Dehumidifiers
The dehumidifier must be operated in an enclosed area to achieve best efficiency. Close all openings to other areas of the building, like windows and doors, to prevent moist air from mixing with the air in the drying area. Keep traffic through doors at a minimum. This forms a closed drying chamber.

Inside the drying chamber, air should circulate freely. Open interior doors and operate TurboDryer airmovers to maintain good airflow to all areas.

Place the dehumidifier where there is no restriction to air flowing through either the inlet or the outlet. Keep away from loose materials like curtains and drapes.

Under normal circumstances, place the dehumidifier in the center of the room. To dry a specific area, place the dehumidifier so the outlet air points at the wet area and the warm, dry air passes across it. The duct should never be closer than three feet from the wall.

The DrizAir Dehumidifier warms the air as it removes its moisture. In smaller rooms this can raise the temperature substantially. Room temperatures of about 68° to 80° F (20° to 27° C) are usually a good condition for drying. Never allow the room temperature to exceed 100° F (38° C), because this could damage the dehumidifier or building contents.

OPERATING INSTRUCTIONS

1. Operate the DrizAir Dehumidifier <u>only</u> in the upright position. If it has been in a horizontal position for any time over a few minutes, let it sit in the upright position for at least 30 minutes before starting. This allows oil to drain back into the compressor, increasing its life.

2. Plug in to a standard outlet with the correct voltage and amperage for the unit. Push the <u>ON</u> switch. Listen for proper operation of fan and compressor before leaving the unit unattended.

07-00265 G   12-02 Copyright 2000, 2002 Dri-Eaz Products, Inc.

FIGURE 4.1   Two versions of the same page showing the effect of format. (*Continued*)

Source: From *Owner's Manual, Dri-Eaz DrizAir® Dehumidifiers, 115-Volt Models: 1200, 2000, 2400,* Dri-Eaz Products, Inc., Burlington, WA, p. 1. Reprinted with permission of Dri-Eaz Products, Inc.

Owner's Manual
Dri-Eaz DrizAir® Dehumidfiers
115-volt models:1200, 2000, 2400

READ AND SAVE THESE INSTRUCTIONS

Use and Operation

⚠ WARNING

FIRE AND ELECTRIC SHOCK HAZARD

Unit must be electrically grounded.

- Insert 3- prong plug on power cord directly into matching grounded receptacle.
- Do not use adaptor

Keep wiring and motor dry.

- Do not operate in standing water
- Do not operate in rain or snow.
- If electrical components become wet, allow them to dry before using.

Read and understand manual before use.

INTRODUCTION

DrizAir dehumidifiers reduce humidity in enclosed structural environments by removing water vapor from the air. With proper use, DrizAir dehumidifiers can help dry wet materials like carpet, carpet pad, floors, walls, building contents and more. Using DrizAir dehumidifiers may also prevent secondary damage caused by high humidity.

For best results, we recommend you use DrizAir dehumidifiers with Dri-Eaz TurboDryer airmovers. The airmovers speed evaporation by lifting moisture into the air.

HOW THE DRIZAIR DEHUMIDIFER WORKS

DrizAir refrigerant dehumidifiers operate by pulling moist air in across a very cold evaporator.core. The moisture condenses (freezes) on the coil. At intervals, the machine will go into defrost mode warming the frost back to water. The water collects in a tray, and leaves the unit through a drain hose or pump.

IMPORTANT: Before moving the unit, make sure there's no water in the pump. See how to use the PURGE function under "How to use the touchpad controller" on page 3.

SETTING UP A DRYING AREA

Use in an enclosed space
You should operate DrizAir dehumidifiers in an enclosed area, as this creates a drying chamber. Close all doors, windows or areas that open to the outside to maximize the units water removal effeciency. Also, keep traffic though the drying chamber to a minimum.

How to position the DrizAir dehumidifier
- Place your DrizAir dehumidifier in the middle of a room away from walls and contents
- Keep it away from anything that could prevent airflow into and out of the unit.

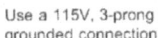
Use a 115V, 3-prong grounded connection

Do not use with an adaptor

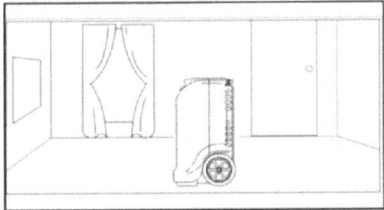
Illustration of proper dehumidifier placement

FIGURE 4.1 (Continued) Two versions of the same page showing the effect of format.

Source: From *Owner's Manual, Dri-Eaz DrizAir® Dehumidifiers, 115-Volt Models: 1200, 2000, 2400,* Dri-Eaz Products, Inc., Burlington, WA, p. 1. Reprinted with permission of Dri-Eaz Products, Inc.

TABLE 4.1
Comparison of Two Formats

Element	Version 1	Version 2
Amount of text on page	More	Less
Amount of white space	Minimal	Lots
Font size	10-point	10-point
Column	Two (even)	Two (uneven)
Justification	Full	Left only
Pictures/illustrations	None	Three

instantly—is the differences that are primarily visual in nature. Table 4.1 shows a comparison of several visual aspects of the two pages.

Some of these attributes interrelate; for example, if you have less text on a page, you will necessarily have more white space or more illustrations (or a combination of both), assuming the page size and font size remain the same. Regardless, taken together, they make a noticeable difference in the "feel" of the page, and may contribute substantially to a user's decision to read a manual or stick it on a shelf to be referred to only if all else fails. Four of these attributes are particularly important:

- Amount of text on a page
- Use of white space
- Font size and style
- Use of columns

We will consider each of these attributes in the next few pages.

WHY LESS IS MORE—STRATEGIES FOR REDUCING TEXT

Less text per page has the effect of making the material seem easier to read. There's a reason that children's books often have only a few sentences per page. Users of manuals are not children, but they are often multitasking when they are using a manual, which reduces the ability to concentrate on the text. Less text per page means fewer opportunities to lose one's place while glancing back and forth from manual to product. And less text per manual makes the manual less expensive to publish and less costly to translate.

How can you reduce text? If you want your manual to contain less text, you can do one of three things (or a combination of them):

- Leave some of the information out.
- Use fewer words to say the same thing.
- Convey some of the information using something other than words.

The first of these has to do with content—what to include and what to leave out—and that has been addressed at length in Chapter 3. This chapter focuses on the other two strategies: concision and alternative communication modes.

Concision

The following poem is one that every writer (at least every technical writer) should post next to the computer. Its origin is uncertain, but its message is sure:

The written word should be clean as bone,
Clear as air, firm as stone.
Two words are not
As good as one.

Sometimes reducing text is a matter of editing for concision. Note that "concise" is not the same as "short." It simply means "no longer than it needs to be." In technical writing, including manual writing, brevity is good, but precision is critical. Do not cut words you need in order to be precise—but you might be surprised how many you can cut without losing meaning. Two common writing habits result in bloated prose that dilutes meaning and expands text: false subjects and unnecessary use of passive voice.

False subjects (also called "expletive openers") refer to words often used at the beginning of sentences that take up space, but do not actually carry meaning. Here are some examples:

- "It is important to note that the threads are reversed."
- "There are two parts to the filter."

In each of these examples, the false subjects serve to push the substantive information toward the end of the sentence while dulling its impact. In the following examples, which is more compelling: A or B?

A	B
"It is important to note that the threads are reversed."	"Important: the threads are reversed."
"There are two parts to the filter."	"The filter has two parts."

Most people prefer B—and B is also shorter. Read over your drafts and ruthlessly cut out deadwood false subjects. Sometimes you may have to rearrange the sentence a little, but in every case, you will end up with stronger, more concise prose.

The second thief of concision is improper use of passive voice. As you no doubt remember from grade-school grammar class, in active voice, the grammatical subject of the sentence is doing the action, while in passive voice, the subject is being acted upon. Here's an example:

Active	Passive
"A toggle switch controls the lights."	"The lights are controlled by a toggle switch."

Passive voice is perfectly correct, and it is useful in some situations, but it always takes more words to say the same thing. Passive voice requires the use of a "helping verb" along with the main verb, and if the writer wants to include the actor, a prepositional phrase (" . . . by a toggle switch") is required as well.

Use passive voice when you want to focus the reader's attention on the receiver of the action (rather than the action itself) or when you don't know the actor. It's rare to have an unknown actor in a manual—in most cases, the reader is supposed to be the actor. That's why you should phrase instructions as commands ("calibrate the instrument" rather than "the instrument must be calibrated"). A command is always active voice—the subject of the sentence is doing the action. In a command, the only possible subject is "you," so it is normally not written, but rather understood. Phrasing an instruction in passive voice can sometimes lead to uncertainty: am I supposed to calibrate the instrument, or is that done at the factory?

Alternatives to Text

Another way to reduce verbiage is to communicate by means of something other than words. Typically in manuals, these other means are charts, graphs, tables, photographs, and illustrations. Any time you use a visual of some kind to convey information, you reduce (and sometimes eliminate) the need for translation, while cutting down on the amount of text. Visuals can actually be more effective than words at communicating certain kinds of information, but to work well, they must be well-designed. Chapter 5 addresses the use of visuals in manuals.

WHITE SPACE IS NOT EMPTY

One of the arguments for reducing text is that it also reduces the cost of the manual—unless the page count remains the same. Sometimes writers are tempted (or pressured) to cram as much text and graphics as possible onto each page in order to cut · production costs—after all, less "empty" space means fewer pages. But eliminating needed white space to cut costs is a false economy: the resulting overcrowded layout will discourage even the most dedicated manual readers. Anyone else will not even try to wade through it. Figure 4.1 illustrates the difference a bit more white space can make. Judicious use of white space provides rest for the eye and gives a document a more open and inviting look.

Properly used white space can also make it physically easier to read a page. Our eyes, like the rest of our bodies, need rest breaks. If these "mini-rests" are not provided to us in the form of white space, our eyes will take them anyway—and we will find ourselves having to reread a sentence because we missed a few words. Also, many people are able to read faster than their minds can follow. White space gives the mind time to assimilate information before going on to the next piece.

But how much is enough? Current practice among manual page designers in the United States calls for roughly a 50/50 division between text and white space, but this preference may vary by culture.[1] Even in the United States, it has changed over time, with older documents much more text-heavy than would be acceptable now. It is also possible to use too much white space. If the white space is so great that it isolates each element and prevents the reader from following the flow of information, it can become a hindrance rather than a help. Strive for a balance between text and white space that both invites readers in and helps them find their way.

White space can help readers understand the structure of a piece of writing. In the same way that headings can show how the manual is organized, white space, if used carefully and consistently, can show how a section of text is put together. Thus, a blank line or two (such as between paragraphs) lets your reader know that you are moving from one unit to another of equal importance. Similarly, indenting a section (widening the white space around it) lets your reader know that you are moving to a smaller organizational division within a single unit. Surrounding an item with white space will also call attention to it, such as setting off a warning from ordinary text for emphasis.

As with any other design element, white space must be used consistently to work properly. In other words, you must always use the same number of blank lines between major divisions in a manual, and the same (smaller) number between minor divisions. You will find it easiest to keep track of this using a table or setting up a template on the computer. Otherwise, you will be having to page or scroll back through completed material to find whether to leave two lines or three between sections.

Careful use of white space may add a tiny bit to the cost of a manual because not every page is crammed corner to corner, but it will help ensure that the manual is used. If the manual sits untouched on the shelf because it looks too hard to read, the entire cost of producing it is wasted.

WHERE ARE MY READING GLASSES?

Aside from the proportion of text to white space, probably the next most critical element in determining whether potential readers see a manual as easy or difficult to read is font size and style.

Font Size: What's the Point?

Type is measured in points, with 72 points to the inch. Before computers, which now do it all electronically, type was printed mechanically. Each line of type was manually set by a typesetter, who aligned individual metal blocks with raised letters on them to spell out a line of type. Point size refers not to the actual letter size, but rather to the height of the metal block on which the raised metal letter sat. Thus, in two different typefaces, the same 12-point letter might be slightly different heights, but the blocks would be the same. Even though no one sets type by hand anymore, the old terms are still often used.

A better measure for comparing two typefaces is x-height. That mathematical-sounding term simply refers to the height of a lowercase letter x. The letter x has the virtue of having a shape that permits easy comparisons—you could lay an imaginary ruler across the tops of the crossed lines that form the x and draw a horizontal line. If you spend a little time with the typefaces available on your computer, you will soon see that the same point size in different typefaces can produce a wide variation of letter sizes. An Arial x, for example, is noticeably larger than a Times New Roman x in the same point size:

X (Arial, 20-point)

X (Times New Roman, 20-point)

Fonts are still commonly referred to in terms of point sizes, so it's useful to have a rough guide for choosing an appropriate size. With 72 points to an inch, the tallest

12-point letter would be a little less than one-sixth of an inch high. The smallest size that can be read without a magnifying glass is 6-point type. Most text is set at least in 8-point type—which still requires reading glasses for those of us over 40. Because manuals may well be used in less-than-ideal conditions, you should go no smaller than 8-point type for anything in the manual. Choosing 10-point or 12-point will greatly increase ease of reading.

The size of the letters themselves is not the only element that affects readability of one point size over another. The other element that comes into play is leading (pronounced "ledding"), another holdover term from hand-typesetting days. Leading refers to the amount of space between lines of text. The smaller the point size, the more relative space is needed. In predigital times, this space was produced by a type-setter adding small strips of lead between lines of type—hence the name. Today, the software typically has an appropriate minimum amount of leading built in, although you can usually override the default setting.

Font Style: A Continuing Debate

Typefaces come in a bewildering variety of styles and sizes, but all English typefaces fall into two broad groups: serif and sans serif. Serifs are the little lines at the end of each stroke in a letter (easily seen at the ends of the strokes creating the x in the Times New Roman example shown earlier). Serif type has these lines, and sans serif type does not (*sans* is French for "without"). Figure 4.2 shows samples of serif and sans serif type.

Which is better? Good question. More than 100 years of research has not been able to produce a definitive answer. Early research purported to show that serif type increased legibility, but a number of the studies were flawed, and all of them may have suffered from a bias toward serif type simply because it was more familiar to most people.[2] Some current studies suggest that users prefer sans serif type for online text.

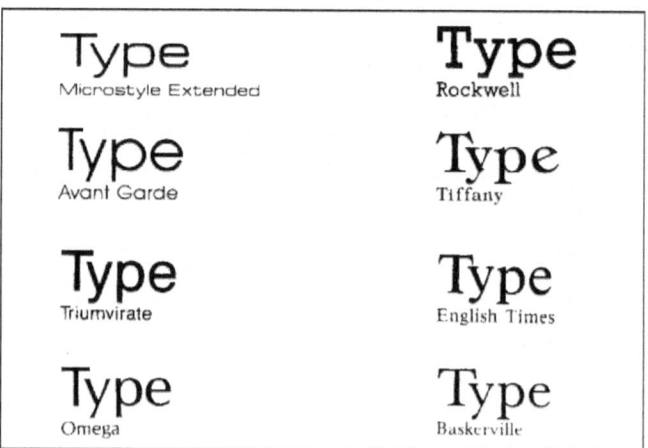

FIGURE 4.2 Examples of Serif (right) and Sans Serif (left). Sans serif typefaces normally use strokes of single weight or width. The Omega typeface shown here is unusual, in that the lines widen out toward the ends—almost suggesting the beginnings of serifs.

Sans serif type has a "modern" look to it, but some people perceive it as being harder to read in large quantities. Although sans serif type may seem harder to read, studies show both styles perform well in actual tests of reader comprehension.[3] Size, as measured by x-height, and the openness of counters (the "holes" in letterforms, such as at the top of a lowercase e) may be more important than whether the typeface has serifs or not. Any of the standard, widely used fonts such as Times Roman or Helvetica can be equally effective in a manual. Avoid excessively ornate fonts or those that resemble cursive writing because they are harder to read and tire the eye more quickly than normal fonts.

THE CASE FOR READABILITY

A final aspect of typography that clearly affects readability is case—whether text is set in a mixture of uppercase and lowercase letters (as in ordinary text) or in all uppercase. Without question, mixed uppercase and lowercase is easier to read, and for a very simple reason. Lowercase letters show greater differences in their shapes than do uppercase letters—all of which are variations on circles, rectangles, and triangles. Thus, it is easier for the eye to distinguish one letter from another in lowercase. A word or phrase (such as in a heading) in all-uppercase type is fine, but never set a whole paragraph in uppercase. Sometimes well-meaning page designers decide that setting the message block of a safety warning in all uppercase letters is a good way to emphasize it. Unfortunately, because that makes the warning less legible (although perhaps more visible), the result is that people will be less likely to read the warning than if it were in ordinary mixed case. A better option for emphasizing text would be to put it in bold type.

HOW MANY COLUMNS SHOULD I USE?

The decision to use one, two, or three columns depends on the interrelationship of several page elements: type size, page size, relationship of visuals to text, and whether your text will be right-justified or not.

Type size, logically, affects line length. A sentence set in 14-point type is significantly longer than the same sentence set in 8-point type, simply because the letters are bigger. Determining the best line length for readability appears to be almost as fraught with controversy as choosing between serif and sans serif typefaces. What is clear from the research is that both very short lines and very long lines are difficult. With a line that is very short, the eye must keep jumping to the next line, words must often be hyphenated, and phrases cannot be seen as a unit, causing fatigue and decreased comprehension. If the line is very long, the eye has too far to travel back to begin the next line and is likely to settle on the wrong line of type. (A corollary is that the longer the line, the more space is needed between lines.) While a longer line length may be acceptable in books and reports, where the reader is expected to read through from beginning to end, it is not a good idea in manuals, where the reader is likely to be looking back and forth between manual and product. The best line length for printed text in manuals seems to be around three inches.

Many manuals use an 8½ × 11-inch page size (more or less). Text set in 10-point type running the full width of the page would be very difficult to read, so page designers have come up with various ways to shorten the line length, primarily by arranging the text in columns. The next set of figures shows some sample pages, all roughly

8½ × 11 inches in the original, with the text laid out in different ways. Figure 4.3 shows a two-column layout with the right margin of the text set ragged, that is, not forming a straight line like the left side. Figure 4.4 shows another two-column layout, this time with the right margin justified (the edge of the text forms a straight vertical line). Figure 4.5 shows a three-column layout, and Figure 4.6 shows a

CARBURETOR ADJUSTMENT

This unit is equipped with a diaphragm-type carburetor that has been carefully calibrated at the factory. In most cases, no further adjustment will be required.

The condition of the air filter is important to the operation of the trimmer. A dirty air filter will restrict the air flow, which upsets the fuel-air mixture in the carburetor. The resulting symptoms are often mistaken for an out-of-adjustment carburetor. Therefore, **check the condition of the air filter before adjusting the carburetor.** Refer to **Air Filter Maintenance.**

If the following conditions are experienced, it may be necessary to adjust the carburetor:

- The engine will not idle
- The engine hesitates or stalls on acceleration
- The loss of engine power that is not corrected by cleaning the air filter and muffler
- The engine operates in an erratic or fuel-rich condition (indicated by excessive exhaust smoke from the muffler).

NOTE: Careless adjustments can seriously damage your unit.

Adjusting the Carburetor

1. Clean the air filter if it is dirty. Refer to Air Filter Maintenance.

2. **Initial Idle Speed Setting:** Turn the idle speed screw **counterclockwise** (Fig. 26) until it *does not contact* the carburetor throttle lever. Now turn the screw **clockwise** until it *begins to move* the throttle lever; then continue turning **2 full turns.**

3. **Initial High Speed and Idle Mixture Setting:** Turn both the high and idle mixture screws **clockwise** (Fig. 26) until they are *lightly* seated. Then turn the screws **counterclockwise 1-1/4 turns.**

4. Start the engine and let it run for a minute.

5. Release the throttle trigger and let the engine idle. If the engine stops, turn the idle speed screw (Fig. 26) **clockwise 1/8 turn** at a time (as required) until the engine idles.

NOTE: Turn the high speed and idle mixture screws finger-tight. Forcing the mixture screws with a screwdriver will damage the screw tip and the seat in the carburetor body.

6. **Final Idle Speed and Mixture Settings:** Adjust the idle speed and mixture for smoothest engine idle.

a. Turn the idle mixture screw (Fig. 26) clockwise until you hear the fastest idle; then turn the screw **counterclockwise 1/8 turn.**

b. Squeeze the throttle trigger. If the engine falters or hesitates as it accelerates, turn the idle mixture screw (Fig. 26) **counterclockwise 1/16 turn** at a time until the engine accelerates rapidly.

c. If the idle speed changes significantly because of steps a and b, readjust the idle speed screw (refer to Step 2).

NOTE: The Bump Head line spool should not rotate when the engine idles.

7. **High Speed Screw Mixture Adjustment:**

a. High speed mixture screw adjustment is not recommended without a precision high speed tachometer.

b. The factory presets the high speed mixture screw at 1-1/4 turns out from the closed position. Your unit should perform well at this setting. If additional adjustment of the high speed mixture is required, contact your local authorized service dealer.

High Speed Mixture Screw

Idle Mixture Screw

Throttle Lever

Idle Speed Screw

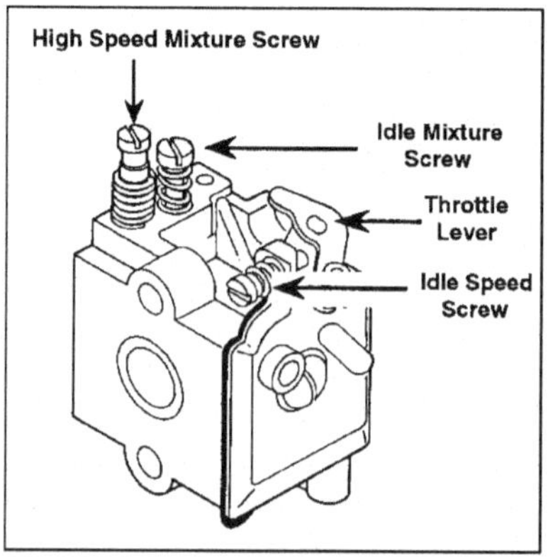

Fig. 26

NOTE: If the carburetor adjustments do not help the unit to run properly, contact your authorized service dealer.

FIGURE 4.3 Two-column layout, set ragged right. In this example, from a manual for a lawn trimmer, the irregular or "ragged" right margin of each column of text gives the manual an informal, open look.

Source: From *Operator's Manual, IDC 500,* Ryobi Outdoor Products, Inc., Chandler, AZ 1998, p. 11. With permission.

Introduction

100 square inches), if each opening communicates with other un-confined areas inside the building. Buildings of unusually tight construction shall have the combustion and ventilation air supplied from outdoors or a freely ventilated attic or crawl space.

If air is supplied from outdoors, directly or through vertical ducts, there must be two openings located as specified above and each must have a minimum net free area of not less than one square inch per 4000 BTUH of the total input rating of all the appliances in the enclosure.

If horizontal ducts are used to communicate with the outdoors, however, each opening must have a minimum net free area of not less than one square inch per 2000 BTUH of the total input rating of all the appliances in the enclosure. If ducts are used, the mini-mum dimension of rectangular air ducts shall be not less than 3 inches.

NOTE: If the openings are to be covered with a protective screen or grill, the net free area of the covering material must be used in determining the size of the openings, as stated above. Protective

screening for the openings MUST NOT be smaller than 1/4 inch mesh to resist clogging by lint or other debris.

Provisions for combustion and ventilation air must comply with ref-erenced codes and standards. See Local Installation Regulations Section on Page 4.

F. CORROSIVE ATMOSPHERES—The water heater should not be installed near an air supply containing halogenated hydrocarbons. For example, the air in beauty shops, drycleaning establishments, photo processing labs, and storage areas for liquid and powdered bleaches or swimpool chemicals often contain such hydrocarbons. The air there maybe safe to breathe, but when it passes through a gas flame, corrosive elements are released that will shorten the life of any gas burning appliance. Propellants from common spray cans or gas leaks from refrigeration equipment are highly corro-sive after passing through a flame. The limited warranty is voided when failure of water heater is due to a corrosive atmosphere. (Reference is made to the limited warranty for complete terms and conditions.)

Installation

1. INSPECT SHIPMENT—Inspect water heater for possible ship-ping damage. Check the marking of the rating plate of the water heater to be certain the type of gas being furnished corresponds to that for which the water heater is equipped.

2. WATER SUPPLY CONNECTIONS— Refer to Fig. 2 for sug-gested typical installation. The installation of unions or flexible copper connectors is recommended on the HOT and COLD water lines, so that the water heater may be easily disconnected for servicing if necessary. The HOT and COLD water connections are clearly marked.

Install a shut-off valve in the cold water line near water heater.

Determine if there is a check valve in the cold water supply line. It may have been installed as a separate component or it may be part of a pressure reducing valve, water meter or water soft-ener.

A check valve located in the cold water inlet line can cause a "closed" water system. A closed system prevents the water, as it is being heated, from expanding back into the cold water sup-ply line. Pressure can build up within the heater causing the re-lief valve to operate during a heating cycle. This excessive op-eration can cause premature failure of the relief valve and pos-sibly the heater itself.

Replacing the relief valve *will not* correct the problem. One method of preventing pressure build-up is to install an expan-sion tank in the cold water supply line between the heater and the check valve. Contact your installing contractor, water supplier or local plumbing inspector on how to control this situation.

IMPORTANT!! This water heater is supplied with one (1) "Cold" (blue tip), and one (1) Hot (red tip) heat trap nipples. They MUST be installed directly into the water heater as shown in Fig. 2. Do not apply heat to the hot or cold heat rap nipple. If sweat connections are used, sweat tubing to

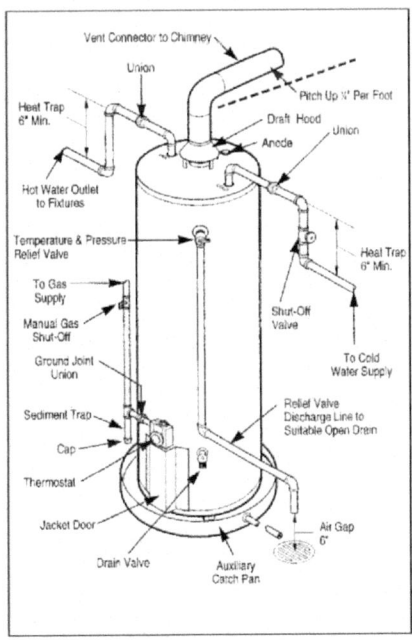

Figure 2 — Typical Installation

FIGURE 4.4   Two-column layout, set fully justified. In this example, from a manual for a water heater, the right margin is justified (straight), giving the page a neat, formal look. Notice, however, that even though the illustration on this page is larger than the one in Figure 4–3, this page appears to have more text on it. Using a fully justified text requires taking care to ensure that each page has sufficient white space to provide rest for the eye and to aid in visual chunk-ing of information.

Source: From *Use and Care Manual, Tri-Power Residential Gas Water Heater,* Rheem Mfg. Co., Montgomery, AL, p. 2. With permission.

USING THE BORDER EDGER ATTACHMENT

Your Mantis Tiller has been designed and built to accept a wide range of Mantis Tiller Attachments to increase its usefulness in your lawn and garden. And, all Mantis Tiller Attachments have been designed for quick and easy attachment to the Tiller or Engine.

The Border Edger

(Item #3222M)

The Most popular attachment, the Border Edger can be used to cut clean, neat edges along walkways, or around trees, shrubs, and garden beds.

The Border Edger has two parts: a wheel and a hardened steel blade, with pointed tines.

How to Install the Border Edger

The following instructions refer to "right" and "left" axles. Assume that you're standing behind your Tiller, as you would for tilling and cultivating.

Some areas of your yard may harbor roots and other underground obstructions. In places like this you'll want to edge your borders shallowly (1" to 2" deep). Here's how to install the Border Edger for shallow edging:

1. First remove your tilling/cultivating tines.

2. Then slide the Edger's wheel onto the right axle.

3. Now slide the Edger blade onto the left axle. The blade's angled face should hit the ground when you spin the blade forward.

4. Insert retaining pins on both left and right axles.

Around walkways and garden beds, you'll want to edge more deeply (3" to 4" deep). Here's how to install the Border Edger for that purpose:

1. Remove the tilling/-cultivating tines.

2. Slide the Edger's blade onto the right axle. The blade's pointed face should hit the ground when you spin the blade forward.

3. Slide the wheel onto the left axle.

4. Insert retaining pins on both sides.

How to Use the Border Edger

1. Position your Mantis Tiller so that the Edger blade is right along the garden edge and the wheel is outside (on the lawn, on the sidewalk, wherever). (Picture 1.)

2. Start your Tiller and pull your Mantis backward along the garden edge. (Picture 2.)

The Border Edger Can Handle Special Projects!

1. Install the Edger for deep edging, as directed above. Then use it to cut sod strips.

2. Edge and weed at the same time! Just attach the Edger blade on one axle and a Tiller tine on the other axle, "Mix and match" blades; don't be afraid to experiment.

Important Note: If you do a lot of edging, you'll appreciate the Mantis Wheel Set (Item #9222M.) It gives you added stability, for even easier handling.

To order the Wheel Set, or any Mantis Attachment, call 1-800-366-6268, Monday through Friday, 9 am to 9 pm, Eastern Time. Ask for the Sales Department.

Picture 1

Picture 2

FIGURE 4.5 Three-column layout. In this example, from a manual for a garden tiller, the text is set in three columns, providing an open look with a good deal of white space. With such a narrow column width, ragged right is a good choice—there's not much room to adjust word or letter spacing as is required for justified text. The column width is ample for the sorts of graphics used here, but would be too small for detailed assembly drawings or charts.

Source: From *Mantis Tiller/Cultivator Owner's Manual*, Mantis Division, Southampton, PA, p. 26. With permission.

Installation Instructions

Step 4

Inlet Hoses

1 Install hoses with the straight end (with filters) fitted to the faucets.

2 Install elbow ends onto washing machine (inlet valves are marked on the back of the machine H=hot, C=cold)

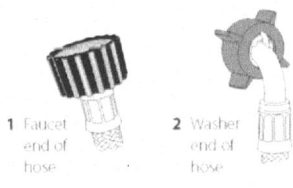

1 Faucet end of hose

2 Washer end of hose

Note:

If there is a cold water supply only, an inlet valve cap (available from Fisher & Paykel) MUST be connected to the hot valve. The cap prevents water leaking from the hot valve.

Step 5

Leveling the Washer

1 Insert the four rubber leveling feet inserts into the feet on the base of the washer. (See diagram 1).

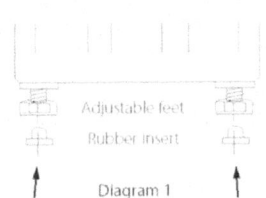

Adjustable feet

Rubber insert

Diagram 1

2 Move the washer into its final position. We suggest a minimum clearance of 1" (25mm) on each side for ease of installation. (See diagram 2).

3 Turn on water and check hose connections for leaks. Check that there are no kinks in the hoses.

1" (25mm)   1" (25mm)

4 It is IMPORTANT to level the washer to ensure proper performance during spin.

Adjust the feet by unscrewing/screwing to make sure the washer is level and cannot rock.

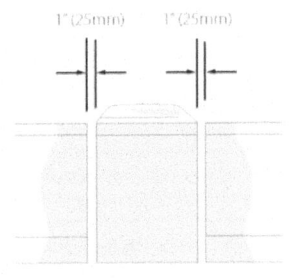

Diagram 2

5 Use the walls and adjacent counter tops as a sight guide to check to see if washer appears level. Readjust the feet if necessary.

6 Open the lid and check the washer tub to see that it sits slightly forward of centre. (See diagram 3).

Incorrect adjustment   Correct adjustment

Diagram 3

8

FIGURE 4.6 Two-column layout with uneven columns. In this example, the wider left column is used for text, and the narrower right column is used for illustrations. Sometimes the narrower column is used for other purposes, such as notes or safety messages.

Source: From User Manual, Eco-Smart GWL15 Washer, Fisher & Paykel, Greenmount, Auckland, New Zealand, p. 8. Source and copyright © Fisher & Paykel Appliances Limited. With permission.

layout with uneven columns. In this format, the wider column is used for text and the narrower one is used (in this instance) for illustrations. On facing pages, the arrangement may be maintained or it may be reversed so that the pages are mirror images of each other. Additionally, the narrow column, which may also be used for headings or other explanatory material, may fall at the inside or the outside of the page.

No one layout is perfect; all have trade-offs. For example, in a two-column format, if the columns are right-justified, the space between them can be narrower without the columns appearing too crowded. But right-justified text (also referred to as "block" text) looks more formal and requires frequent hyphenation. In addition, justifying the right margin means that the spacing between letters and words must be adjusted to make all the lines come out the same length. The narrower the column, the harder this is to do well. We have all seen dreadful examples in newspaper columns where one long word has been s t r e t c h e d across a whole column, and in the process has become nearly unreadable. Before you opt for right-justified margins, be sure to check out how sophisticated your publishing software is. Badly spaced words and letters can make text very difficult and unpleasant to read.

Whatever column arrangement you decide on, stick with it for the whole manual. It is quite disconcerting to the reader to turn a page and find that the format has suddenly switched from two columns on a page to three. Similarly, be careful that your illustrations do not protrude into the space between columns of text. They should be sized to fit within the column or to extend over two or more full columns. If the illustrations are allowed to "leak" into the "gutter" between columns, the page will look visually disorganized and difficult to read. Remember, the object in designing a format is to make the pages look inviting. A clean, consistent, ordered layout gives the impression that the material is well-organized and easy to read.

HEADINGS: A MAP FOR YOUR READER

When you are driving to a location you've never visited, it's comforting to have a map. Navigation software can help you find your way from Point A to Point B, but a physical map allows you to see your current location in relation to others. Similarly, proper sentence structure and tight transitions help you find your way through a paragraph, but a good heading structure functions more like a map. Headings give the reader information about both sequence and hierarchy of information.

WHAT COMES NEXT? SIGNALING SEQUENCE

Even if you provide a well-designed table of contents for your manual, many people will simply flip through the pages to find what they're looking for. What they are actually doing is skimming through the headings—not the text itself, at least not until they've located the section of interest. The headings act as labels identifying the content of a particular section of text. As the reader turns the pages, the logical sequence of information is revealed. Chapter 3 discusses how to organize information; your heading structure will reveal that organization to the reader. For example, suppose you have decided that the following sequence of information would be best for your manual:

- Safety
- Application and Special Uses
- Installation
- Operation
- Maintenance
- Storage
- Troubleshooting

The general headings of your manual should more or less match that list. Very quickly your reader will recognize that you have chosen to sequence information chronologically, and will have a good idea of where to start looking for setup instructions, or information on changing the oil, or calibrating the probe, or whatever he or she needs. As you develop the outline for the major sections in your manual, you also determine the major headings.

SIGNALING HIERARCHY: PUTTING THE PIECES TOGETHER

Just as a cartographer making a geographical map uses color, line weight, etc., to make a four-lane freeway look different from a gravel country road, a page designer can use typography to differentiate between major sections and minor ones. Headings not only label the main content of the section to follow, but also provide cues to the relative importance of various kinds of information within a section and between sections. The reader picks up these cues (often unconsciously) in responding to print size, typeface, uppercase or lowercase letters, and position of the heading on the page. Individuals may disagree here and there, but in general, readers will interpret the following from typography and position:

1. Highest level of organization or most important information shown by one or more of these:
   a. Larger type
   b. Bold print
   c. All uppercase letters
   d. Centered on the page
2. Lower level of organization shown by one or more of these:
   a. Smaller type
   b. Finer print
   c. Mix of uppercase and lowercase letters, or all lowercase
   d. Flush left or indented
3. Special emphasis shown by one or more of these:
   a. Italics, underline, or bold
   b. Boxes
   c. Color

MAKING IT WORK

The best heading structure is an accurate reflection of text. In other words, it tells what comes first, what second, and so on, and it tells how those sections are arranged into major and minor divisions. Be sure to use enough headings. Most writers use too few. Instead of just labeling the major sections of a chapter, consider adding subheadings to point out the smaller divisions. Remember that, as the writer, you are familiar with what is contained in the manual you are writing—you know where the parts are and what is covered in each. Your readers are looking at the manual for the first time; without sufficient headings to help them find their way, the manual will look like a bewildering sea of prose.

On the other hand, be sure that your heading structure is not overly complex. Limit yourself to no more than three levels of headings—more than three becomes too confusing for the reader. Make sure that the type size and style that you have selected for the different levels of headings signal the shifts in an intuitive way. For example, as noted earlier, we automatically assume that larger type identifies a major section, smaller type a subsection. Similarly, bold type of a certain size signals a more major division than normal type of the same size. A straightforward, intuitive heading design will help guide your readers without their even being aware of it.

How do you decide what your headings should be? The simplest procedure is to use the outline for the text to guide you. Outlines show both the sequence of information and the hierarchy. Using the outline as a guide for the heading structure will ensure that your headings consistently and accurately reflect the organization of the material. The publication software your company uses may allow you to generate appropriately formatted headings automatically from an outline. The low-tech method of going through your outline with different colors of highlighter pens for the different levels of headings works just as well!

Figure 4.7 shows the power of print size, uppercase and lowercase, boxes, lists, and italics to signal structure. When manual users see a page like this, they immediately identify these cues to the organization:

- The page addresses two major topics: "Storage" and "Starting After Long Term Storage."
- The first section includes two minor topics: "Temporary Storage" and "Long Term Storage."
- The bulleted list conveys important information—but the information does not have to appear in a specific order.
- The numbered lists convey information that does need to follow a specific sequence.
- The boxed text, which is also in bold type and accompanied by a safety-alert symbol and a signal word ("WARNING"), conveys important safety information.
- The italicized note conveys information that requires special emphasis.

Operating the Tractor

STORAGE

Temporary Storage (30 Days Or Less)

Remember, the fuel tank will still contain some gasoline, so never store the unit indoors or in any other area where fuel vapor could travel to any ignition source. Fuel vapor is also toxic if inhaled, so never store the unit in any structure used for human or animal habitation.

Here is a checklist of things to do when storing your unit temporarily or in between uses:

- Keep the unit in an area where children will not come into contact with it. If there's any chance of unauthorized use, remove the keys and disconnect the spark plug wires.
- If the unit can't be stored on a reasonably level surface, chock the wheels.
- Clean all grass and dirt from the mower.

NOTE: If storing your tractor between winter snow removal jobs in a cold area, we suggest that you fill the fuel tank at the completion of each job to prevent water condensation in the fuel tank. Wait for engine to cool before filling tank.

Long Term Storage (Longer Than 30 Days)

Before you store your unit for the off-season, read the Maintenance and Storage instructions in the Safety Rules section, then perform the following steps:

1. Drain crankcase oil and refill with a grade of oil that will be required when unit is used again.
2. Prepare the mower deck for storage as follows:
   a. Remove mower deck from the unit.
   b. Clean underside of mower deck.
   c. Coat all bare metal surfaces with paint or light coat of oil to prevent rusting.
3. Clean external surfaces and engine.
4. Prepare engine for storage. See engine owner's manual.
5. Clean any dirt or grass from cylinder head cooling fins, engine housing and air cleaner element.
6. Cover air cleaner and exhaust outlet tightly with plastic or other waterproof material to keep out moisture, dirt and insects.
7. Completely grease and oil unit as outlined in the Regular Maintenance section.
8. Clean up unit and apply paint or rust preventative to any areas where paint is chipped or damaged.
9. Be sure the battery is filled to the proper level with water and is fully charged. Battery life will be increased if it is removed, put in a cool, dry place and fully charged about once a month. If battery is left in unit, disconnect the negative cable.

⚠ WARNING

Never store the unit, with gasoline in engine or fuel tank, in a heated shelter or in enclosed, poorly ventilated enclosures. Gasoline fumes may reach an open flame, spark or pilot light (such as a furnace, water heater, clothes dryer, etc.) and cause an explosion.

Handle gasoline carefully. It is highly flammable and careless use could result in serious fire damage to your person or property.

Drain fuel into an approved container outdoors away from open flame or sparks.

10. Drain fuel system completely or add a gasoline stabilizer to the fuel system. If you have chosen to use a fuel stabilizer and have not drained the fuel system, follow all safety instructions and storage precautions in this manual to prevent the possibility of fire from the ignition of gasoline fumes. Remember, gasoline fumes can travel to distant sources of ignition and ignite, causing risk of explosion and fire.

NOTE: Gasoline, if permitted to stand unused for extended periods (30 days or more), may develop gummy deposits which can adversely affect the engine carburetor and cause engine malfunction. To avoid this condition, add a gasoline stabilizer to the fuel tank and run the engine a few minutes, or drain all fuel from the unit before placing it in storage.

STARTING AFTER LONG TERM STORAGE

Before starting the unit after it has been stored for a long period of time, perform the following steps:

1. Remove any blocks from under the unit.
2. Install the battery if it was removed.
3. Unplug the exhaust outlet and air cleaner.
4. Fill the fuel tank with fresh gasoline. See engine manual for recommendations.
5. See engine owner's manual and follow all instructions for preparing engine after storage.
6. Check crankcase oil level and add proper oil if necessary. If any condensation has developed during storage, drain crankcase oil and refill.
7. Inflate tires to proper pressure. Check fluid levels.
8. Start the engine and let it run slowly. DO NOT run at high speed immediately after starting. Be sure to run engine only outdoors or in well ventilated area.

FIGURE 4.7   Use of formatting elements to signal structure. Notice how the use of upper-case and lowercase, varied print size, lists, and italics signal the organizational structure automatically.

Source: From *Operator's Manual, Regent/2500/500 Series, Rev 9/2002,* Simplicity Manufacturing, Inc., Port Washington, WI, p. 16. Briggs and Stratton Power products Group, LLC. Used with permission.

OTHER SIGNPOSTS FOR YOUR READER

In addition to the heading structure, some manuals include other cues to help readers find their way. These include headers and footers, numbering systems, and physical organizers.

HEADERS AND FOOTERS

One kind of signpost that the reader can use to find his or her way to the information needed is the headers and footers at the top and bottom of the pages. Headers (also known as "running heads") are headings at the top of a page that tell the reader what part of the book he or she is in. For example, it is common in a manual to show the chapter title at the top of the left-hand page and a subhead from within the text on the right-hand page. Less commonly, this information is given at the bottom of the page instead of the top, in which case the labels are called footers (not, however, "running feet!"). More commonly, footers, if used at all, are used for such information as revision number, series, or date. Running heads allow the user to flip through the pages quickly to find the relevant section, and then shift to using headings within the text to locate the precise information needed. Figure 4.7 has a running head ("Operating the Tractor") identifying the larger unit of the manual that includes that page.

NUMBERING SYSTEMS

Some manuals number each paragraph. Various systems are used; perhaps the most familiar is that used in the military. In this system, hierarchy is indicated by numbers divided by periods. Thus, paragraph 1.2 is of a higher level than paragraph 1.2.1. In outline form, it might look like this:

    1.0 Chapter One
        1.1 First subtopic
        1.2 Second subtopic
            1.2.1 First sub-subtopic
            1.2.2 Second sub-subtopic
    2.0 Chapter Two . . .

The advantages of such a system are that it is easy for the reader to see the organizational structure of a manual and that it is possible to direct a reader to a particular paragraph (rather than just a page) for specific information. Further, if a company has several similar products, common "boilerplate" information may be easily catalogued on the computer by giving each paragraph a unique address, making it possible to assemble relevant information to be shared among various products quickly and easily.

The system has some disadvantages, however. It is a bit of a distraction to the reader, adding a level of clutter to page layout that may or may not outweigh the usefulness. Whenever the manual is revised, even in a minor way, the paragraphs must be renumbered, particularly if information is inserted rather than deleted. It may get quite cumbersome if there are many levels to the outline. Finally, it may be confusing to the reader when combined with other writing strategies such as using numbered lists for explaining procedures.

Whatever system is used, it must show hierarchy as well as sequence. Some years ago, certain UN documents used a purely sequential numbering system: paragraphs would be numbered in order from the beginning to the end of the document. While such a system allows references to specific paragraphs, it does not help the reader see the structure at all, and any revision requires renumbering the entire document.

When a paragraph numbering system is used, the index (and sometimes the table of contents) generally refers the reader to the paragraph number rather than the page number associated with the listed topic. As long as this is clearly indicated, it should pose no problem, although it is a little more difficult to flip through a manual to find a specific paragraph number than to find a specific page number—simply because the page numbers are always in the same location on the page.

Physical Organizers

Some manuals include physical organizers such as tabbed dividers that literally separate different sections of the manual. The section name may be shown on the tab. Sometimes these divisions are signaled by different colored pages. When different sections of the manual are separated this way, they often also signal the section change by the page numbering. For example, instead of numbering pages sequentially from beginning to end, they may be divided by chapter: the pages in Chapter Five, for example, would be shown as 5–1, 5–2, 5–3, and so on.

Physical organizers can be helpful in very long manuals. Auto service manuals, for example, often have tabs dividing the section on engine repair from the section on the cooling system or transmission. If the manual is too short, these dividers may be more of a hindrance than a help, making it cumbersome to use, much as advertising inserts printed on stiff stock make it difficult to flip through a magazine.

PAVING THE ROAD FOR YOUR READERS

You've chosen appropriate content; you've organized your information logically; you've invited the reader in with an attractive format; and you've created a headings "map" and put up signposts to help the reader find his or her way. What's left? To push the map analogy a little farther, a detailed map and prominent road signs are all well and good, but the road itself can still be full of ruts or potholes. What can you do to smooth out the path and maybe even lay down some blacktop for your readers? When you drive on a newly resurfaced highway, you don't notice the pavement—precisely because it's smooth and presents no obstacles to easy passage. Similarly, good technical writing puts up no barriers for the reader.

Accomplishing that involves four fundamental tasks:

- Writing from the user's point of view.
- Writing correctly.
- Writing precisely (getting rid of weasel words).
- Checking your work (usability testing) and incorporating feedback.

These are simple—but not always easy.

WRITING FROM THE USER'S POINT OF VIEW

Chapter 2 and Chapter 3 address finding out who your users are and choosing the content and organizational strategies that would best serve them. One more aspect requires you to put yourself in your users' shoes: choosing words and writing individual sentences. Once again, the most difficult aspect is looking at your product as a first-time user would. Because of your familiarity with the product, you will take for granted knowledge that the first-time user will not have.

The most easily noticeable aspect that may prove difficult for a first-time user is terminology. You may know that the "tubular end closures" are the little plastic caps that fit over the bottoms of the legs on the charcoal grill, but the first-time buyer may not. You may know that "dressing the blade" means sharpening it, but the first-time user may not. And you may know that the "air hose coupler" is the same as the "air supply coupling," but the first-time user may not. These examples highlight three language problems that often confound first-time users of a product:

1. Unfamiliar terms
2. Jargon or in-house terms
3. Inconsistent terms

Chapter 3 touches on these, but it's worth a second look. How can you avoid these pitfalls?

Unfamiliar Terms

The best way to handle unfamiliar terminology is to decide on appropriate terms for the various parts of your product—and the terms do not have to be exactly what the engineers call them. The engineer may call it an "electronic actuator," but you can call it an "on—off switch."* If the location and function of a part would not be immediately obvious to anyone (be pessimistic), be sure that you include a parts diagram with all referenced parts clearly labeled. Unfamiliar terminology presents a significant obstacle for readers. It's hard to understand instructions when you don't know what the instructions are talking about. In addition, the same part may be called different things by different manufacturers, so even if a user is familiar with the general product category, he or she may not be familiar with your company's terms. For that reason, it is best to use common terms whenever possible. (They are also more likely to be translated correctly.)

Jargon or In-House Terms

The best approach for jargon or in-house terms is to eliminate them. Idiosyncratic terms are unlikely to be understood by even experienced users and certainly will not be understood by first-time users. They are not likely to be translated correctly. The trick is to recognize them—when you work in the company where these terms originated, after a while they just become second nature. An outside reader can be helpful. If you eliminate in-house terms, what should you use in their place? Either

---

* Some years ago, an engineer in one of my technical writing classes, describing his tent being struck by lightning during a camping trip, said that the problem was that "the tent had aluminum beam and columns." I noted that most people call those items tent poles.

substitute a term that is already common in the industry, or name the part using ordinary words. To give an example, suppose your company manufactures winches. For some reason lost to history, your in-house term for the movable part that locks the winch by wedging against the teeth of the sprocket to keep it from turning is called the "camjam." That term would not be understandable either to a first-time user or an experienced winch user. Nor would it be readily translated. You can either substitute the more widely known term ("pawl") or simply name it something more identifiable ("sprocket lock" or "sprocket stop," perhaps). Either option would give the reader information as to location and function, which "camjam" does not.

Inconsistent Terms

Inconsistent terms pose a particularly difficult problem. For example, if you call the same part a "flange" in one part of the manual and a "collar" in another, all your readers—experienced or not—are likely to be confused. Using two different terms suggests that you are talking about two different parts. Readers will lose time and patience trying to figure out whether you are talking about the same item or not. Even worse, if your manual is translated, the translator is very likely to use different words for each term (they aren't the same, after all). The words chosen may have very different connotations than the English words "flange" and "collar." For example, an online translation engine offers *pestaña* as a Spanish translation for "flange," and *cuello* for "collar." *Pestaña* literally means "eyelash," and *cuello* literally means "the collar on a piece of clothing." If your manual were later translated from Spanish into yet another language, you can imagine how far afield the meanings might become. Certainly, a good technical translator will do a better job than plugging individual words into an online dictionary, but the essential problem remains—if the translator does not realize these two different terms actually refer to the same part, he or she will probably use different words when translating them.

WRITING CORRECTLY

Writing correctly means using correct grammar, spelling, and punctuation. Certainly, correct grammar and punctuation are essential for ensuring that your work looks professional, but they are important for another reason as well. Errors in the "mechanics" of writing are at best distracting, and at worst misleading. Remember, you want your writing to be completely transparent—unnoticeable. Compare the two short passages shown in Example 4.1.

Example 4.1: Errors Prevent Smooth Reading

Passage A

A semiautomatic firearm goes through four stages in a complete cycle of operation:

1. The firing pin (or striker) strikes the percussion cap of the cartridge, causing the primer to explode and ignite the powder.
2. As the powder rapidly burns, the rising pressure forces the bullet out the muzzle of the pistol and simultaneously drives the slide to the rear, putting the mainspring into tension.

3. As the slide travels to the rear, the extractor hooks the edge of the spent casing, pulling it rearward until it strikes the offset ejector, which tips the spent casing out the ejection port.
4. When the slide reaches the end of its rearward travel, the mainspring begins to contract, pulling the slide forward. As the slide moves forward, it recocks the hammer (or resets the striker) and pushes the next round from the magazine up the feed ramp and into the chamber, readying the next shot.

Passage B

A semiautomatic firarm goes through four stages in a complete cycle of operation:

1. The firing pin (or striker) strikes the percussion cap of the cartridge, causing the primer to explode and ignite the powder.
2. As the powder rapidly burns, the rising pressure, forces the bullet out the muzzle of the pistol and simultaneously drives the slide to the rear, putting the mainspring into tension.
3. As the slide travels to the rear, the extractor hook the edge of the spent casing, pulling it rearward until it strikes the offset ejector, which tips the spent casing out the ejection port.
4. When the slide reaches the end of its rearward travel, the mainspring begins to contract, pulling the slide foward. As the slide moves forward, it recocks the hammer (or resets the striker) and pushes the next round from the magazine up the feed ramp and into the chamber, readying the next shot.

Did you notice the errors? Passage B includes two spelling errors ("firarm" and "foward"), one extra comma (after pressure in #2), and one verb form error ("hook" should be "hooks" in #3). These are relatively minor errors that would not interfere with understanding the meaning of the passage at all—but they are distracting, and therefore slow the reading a little. Even for a just a moment, the reader's attention is off the topic of firearm operation and on the language itself.

Even worse are errors that lead to confusion. Without properly placed hyphens, does the phrase "ten foot long pieces" mean ten pieces, each one foot long, or several pieces, each ten feet long? In the following example, what does "it" refer to?

To replace the control lever, loosen the jam nut and unscrew the shaft. Be sure to retighten it.

These kinds of errors may be difficult to catch, because whether the pieces are intended to be one foot or ten feet long and exactly what needs to be retightened. That is why it is best not to proofread your own writing. If you must proofread your own work, at least let some time pass between the writing and the proofreading— you'll spot more errors.

The more you know about grammar, the more you can use sentence structure to convey emphasis and signal importance, without having to explicitly state the relationships.

Consider Example 4.2. Changing Emphasis Through Sentence Structure. The information in all four constructions is exactly the same. What changes from one to the next is the relative importance of the two parts. In the first construction, the fact that the samples were taken daily and the fact that they showed increasing ozone levels are given equal weight. Simply by manipulating the grammar—demoting "daily" from a full sentence to an independent clause, to a phrase, to a single word—the emphasis shifts, making the increasing ozone levels relatively more and more important.

Example 4.2: Changing Emphasis through Sentence Structure

The samples were taken daily. They showed increasing ozone levels.
The samples, which were taken daily, showed increasing ozone levels.
The samples, taken daily, showed increasing ozone levels.
The daily samples showed increasing ozone levels.

Most writers make these kinds of syntactical choices almost as unconsciously as readers respond to them. But sometimes things go awry, especially if the writer is interrupted mid-sentence by a ringing phone or a question from a colleague. When you read over your work, if it seems a little "off," check to make sure that the grammar is sending the right message.

WRITING PRECISELY: GETTING RID OF WEASEL WORDS

Much like false subjects, weasel words give the impression of substance and precision where none in fact exists. Here are a few examples:

- Be sure to clean the filter regularly.
- Lubricate frequently with a small amount of oil.
- Use with adequate ventilation.
- Tighten the belt as needed.

How often is "regularly" or "frequently"—daily or once a month? How much is a small amount? What constitutes adequate ventilation—a window open or a supplied-air respirator or something in between? What does "as needed" mean? How does one know if the belt needs to be tightened? These phrases appear to give the reader guidance, but do not actually provide useful information.

A better approach is to provide specific time intervals or specifications or indicators that maintenance is needed. For example:

- Clean the filter when it becomes visibly dirty.
- Lubricate weekly with two drops of light machine oil.
- Use outdoors only or wear a supplied-air respirator.
- Tighten the belt when it starts to slip or can be deflected more than one-quarter inch midway between the pulleys.

It takes more work to be precise, but doing so will help your users operate and maintain your company's products properly and safely and will keep your readers from getting frustrated.

A different kind of weasel word is one in which usage has sucked out the meaning, much as an actual weasel will pierce an egg and suck out the contents, leaving the shell apparently intact. Language is an ever-evolving entity, and sometimes words change meaning over time. A good example is the word "bimonthly." If you ask any group of people whether "bimonthly" means every two months or twice a month, you will find the group divided. I have done this repeatedly with groups of technical writers (certainly more word-conscious than the average person) with consistent results—between one-third and one-half of the group think it means one thing and the rest think it means the other. (Which definition garners the majority vote varies.)

What's does the word mean? If you go to a current dictionary, you'll find both definitions. Historically, "bimonthly" meant every two months, in contrast to "semimonthly," which meant twice a month. Dictionaries now report how words are used rather than serve to enforce proper usage (they are now descriptive rather than prescriptive). What's the bottom line? Simply this: as a technical writer, you cannot afford to use the word "bimonthly," because you cannot ensure that the reader will understand it the same way you do. The better option is simply to say "every two months" or "twice a month." In this situation, three words *are* better than one—because the one is no longer precise.

MAKING THE MANUAL PHYSICALLY EASY TO USE

Most of this chapter has addressed how to make the manual look inviting and accessible and how to make it easy to navigate and read. Format and font are important choices, but they are not the only choices manual designers must make. Unless you are creating online documentation that will appear in Web-page form only, you have other decisions to make: For example, what is the appropriate page size? Will the manual include different languages? Will it need to be updated frequently? What kind of paper should it be printed on? What kind of binding should it have? All these questions address design choices that will affect how easy it is for the customer to use the manual and how easy it is for the company to revise it.

Size

What is the best size for a manual? That depends on the type of product and how the manual will be used. An 8½ × 11-inch three-ring binder would be an awkward size for a manual for a digital camera or a handheld GPS unit—you want something small to fit in the case. On the other hand, it would be a fine size to use on a workbench in a garage. A relatively simple product, such as a non-motorized push mower, does not need a complicated manual—a few sheets of paper stapled together will work just fine. A complex computer-controlled printing press, however, will require a much thicker book—or books—to convey the needed information. As always, form follows function. Think about how your customer will use the manual, and size it accordingly. Except for small consumer products, manufacturers are increasingly opting for a standard 8½ × 11-inch page size.

Also consider how the company would like to have the manual used. It may be important in reducing liability exposure to have it convenient to keep the manual with the product. Sizing an operator manual to fit in a compartment built into an industrial machine might help ensure that the machine operator has access to the manual. On the other hand, the three-ring binder with letter-size pages is the standard for industrial reference manuals. In general, it is best to use common sizes: oversized manuals won't fit in a bookcase and small, odd-sized manuals are easy to lose.

PAPER

The paper used in a manual must be durable enough to last as long as the document will be used. If a manual is likely to be read once and discarded, the paper need not be of the best quality. On the other hand, if the manual will be referred to again and again, the paper needs to be durable. Another consideration is that the flimsier the paper, the more likely it is that the printing on one side of the page will bleed through, or be visible from the other side. Bleed-through makes reading more difficult.

The paper used for a manual must also not be too porous. Porosity affects how the paper accepts ink and thus how sharp photographs and illustrations will appear. If you have ever tried to write on a paper napkin with a felt-tip marker, you have experienced the tendency of porous paper to allow ink to bleed. If the ink can bleed, lines become fuzzy, and fine detail is lost. Porous paper will also readily accept other substances, like oil and dirt. If your manual is likely to be used under dirty conditions, you should choose a harder-surfaced paper, possibly even a coated stock (although this option is quite expensive).

BINDING

A good binding holds the pages of a manual together so that they can be easily read. Consider these factors:

- Will the pages lie flat? Nothing is more irritating than trying to do a procedure requiring both hands and frequent glances at the instructions, only to find that the manual flops shut each time you let go.
- Will the pages begin to fall out after hard use? This is a common problem with "perfect" bindings and with comb bindings, though rarely with stapled, stitched, or spiral bindings.
- Will frequent additions or corrections be sent to owners? If the manual is likely to be updated often, it might be a good idea to put it in a ring binder, so that outdated pages can easily be replaced with new ones. If the binding requires that holes be punched in pages, make sure that margins are wide enough to allow this to happen without losing part of the text or illustrations. An alternative for occasional changes is to send adhesive-backed pages to owners. These pages can be stuck onto the existing page in a bound manual. It is certainly a more expensive option, but for some vital information it might be worth the cost.

COVER

The cover of a manual influences both how likely it is that the manual will be used and how easy it is to use. An attractive cover invites readers and presents a good image of the company. Remember, the manual offers an avenue for the company to show its customers that it cares about them. An attractive manual cover is the first step. Most companies have a standard cover format for their manuals, including information about the model and often a picture of the product. A good cover should also include the company's address and telephone or fax number and Web address (the back cover is fine for this).

The cover of a manual must also be stout enough to protect the inside pages under the expected conditions of use. Clearly, if the manual will be used in a harsh environment (rain, dirt, oil, etc.), it should be made of coated stock, and possibly even of vinyl. On the other hand, a manual that will receive only occasional use in a clean office may have a cover of the same paper as the inside pages.

SUMMARY

Knowing your audience and organizing information are not enough in today's competitive world. If your manual does not look attractive and inviting and does not help readers find what they need, it won't get used. Just as marketers know that the packaging of a product can be as important as the quality of the product itself to make the sale, the packaging of a manual can make the difference between one that is used and one that is left on the shelf.

Good page design, including an open, inviting look with ample white space and a readable font, will help users make the decision to give the manual a try. Page layouts that look organized and intentional will appear accessible, while confusing layouts that switch from two-column to three-column or have visuals placed haphazardly on the page will put potential readers off. Once readers make the decision to try the manual, a well-designed heading structure can guide them through the content. Headers and footers, numbering systems, and physical organizers can provide additional help to readers so that they can find what they need right away, without wasting time looking.

Equally important is making sure that the text itself puts up no barriers to easy reading and comprehension. Making sure that terminology is logical and consistent, sentences are grammatically correct, and syntax properly conveys the relationship of ideas and information are all important to easy reading. Cutting out deadwood, such as false subjects, unneeded passive voice constructions, and weasel words, will ensure that your writing is tight, concise, and precise.

Finally, designing the physical aspects of the manual—size, paper, type of binding, and cover—to work well in the anticipated conditions of use will help make your manual more usable, and therefore more likely to be used.

Many people, perhaps especially those who work with mechanical products, are visual learners. For those people, a picture is truly worth a thousand words—if it's a good picture. The next chapter explores how you can use visuals to convey information sometimes even more effectively than with words.

CHECKLIST: MAKE IT EASY FOR YOUR READERS

- Is the manual's page layout attractive—not too crowded or jumbled? Does it look organized and purposeful?
- Is there enough white space?
- Is the font large enough and easy to read?
- Does the manual have a logical heading structure? Have I used enough headings?
- Are my sentences written for the user rather than for the engineer?
- Are the grammar and punctuation correct?
- Have I pruned out excess verbiage and weasel words?
- Is the cover of the manual attractive and informative?
- Is the manual a convenient size?
- Will the binding stand up to the conditions of use?

NOTES

1. Mary Lou Fisk, "People, Proxemics, and Possibilities for Technical Writing," IEEE *Transactions on Professional Communication* 35, no. 3 (1992): 176–182.
2. For an interesting paper on the legibility of serif vs. sans serif fonts, see www.sciencedirect.com/science/article/pii/S0042698905003007?via%3Dihub, accessed January 02, 2019.
3. See, for example, Rudi W. De Lange, Hendry I. Esterhuizen, and Derek Beatty, "Performance Differences between Times and Helvetica in a Reading Task," *Electronic Publishing* 6, no. 3 (1993): 241–248.

5 A Picture Is Worth ... It Depends

OVERVIEW

For decades now, the average IQ in industrialized nations has been increasing—and the increase is greatest in the areas of visual intelligence.[1] One possible explanation of this phenomenon is the increasingly richer visual environment we all encounter every day.[2] One hundred years ago, the visual stimuli encountered by most people included the natural physical world, paintings and sculpture, and perhaps an occasional map, diagram, or chart. Today, we spend many hours a day engaged with visual media, whether it be television, YouTube videos, webinars, PowerPoint® presentations, or video games. Most people today get their news from the Internet or television rather than from a newspaper. We have become increasingly adept at processing visual information. Yet many manuals contain far more textual material than visual.

Technical writers are typically more comfortable with words than with illustrations. If they were trained as technical writers, they may have had a class or two on graphics or web design, but seldom more than that. If they were not trained as technical writers, but "fell into" the profession (and that remains the career path for a surprising number of technical writers), they probably got into documentation because they had a way with words. Increasingly, expertise with visual communication is becoming a critical skill.

For most users, the graphics in a manual play a big part in creating a first impression—which may determine whether the manual is used at all. Most users, confronted with a manual for the first time, will leaf through the "pictures" before they read the words. Often the graphics—photos, drawings, charts, and tables—determine whether the manual looks inviting or forbidding. Most people will try to figure out a procedure from the pictures first and read the text only as a last resort. Readers will not only see the pictures before the text, they will also usually expect them to be more accessible than the text. Graphics can break up large blocks of text and give the manual a more open, "friendly" feel. By contrast, a text-only manual looks extremely daunting to most readers, even if the font is large enough for easy reading and there is adequate white space. Sharp, clear photos and illustrations convey that the company cares enough about the customer to produce a top-quality, helpful manual. On the other hand, fuzzy pictures and cluttered drawings leave the reader with the feeling that the manual was an afterthought at best. While it is often said that "a picture is worth a thousand words," that is true only if the picture is well-designed and well-executed.

Graphics, of course, do much more than just make an impression. They are a vital part of the communication package, and in our increasingly visually oriented society, may become the primary communication channel for some purposes and some audiences. Graphics work better than words to help a user identify parts on a product or learn to do a procedure. For readers who do not read well or do not know English, graphics may be the only source of information about the product. Even in a translated manual, the graphics remain critical—they are not subject to the vagaries of poor translation. Indeed, because they don't need to be translated, including graphics will reduce the cost of producing manuals for foreign markets.

This chapter provides an introduction to effective use and design of graphics in manuals. The best graphics result when manual designers plan them from the very beginning, right along with the text. When graphics are a part of the plan from the start, they can be integrated seamlessly with the written portions so that text and art work together to convey the needed information to the user. Good design of graphics follows the same basic tenets of good textual design: graphics must be easy to use and must make important information stand out. As always, form follows function—you should include only those visual elements that help advance information transfer.

WHEN TO USE GRAPHICS

To this point, much of this book has discussed how to plan and write effective instructions. Well-written descriptions of products and explanations of procedures are important ways to convey needed information to users. But they are not the only ways to do so—and sometimes not even the best ways. Clear, readable instructions and descriptions are necessary, but clear visuals are vital.

Some things are simply easier to grasp when they are presented graphically rather than in words. But how do you know when to use a visual rather than words alone? Ask yourself four questions:

- Is there a picture in my mind?
- Do I have a lot of numbers to convey?
- Do I have a procedure to explain?
- Will my manual be translated?

Whenever the answer is yes, consider using a visual.

Is There a Picture in Your Mind?

Often when you are writing text for a manual, you will find that you are visualizing some aspect of the product. If you have a picture in your mind, you should give your user a picture as well. Two situations, in particular, demand visuals:

- Identifying parts
- Explaining spatial relationships

Identifying Parts

All manual users should rely on the pictures to help make unfamiliar part names clear: when told to "tap leg closures firmly until well seated," even the most word-oriented new barbecue-grill owner will look at the drawing to find (with relief) that the leg closures are merely the little plastic caps that go over the ends of the grill's tubular metal legs. Graphics are used for parts identification throughout a manual, from the initial setup chapter to the parts catalog at the end. Words absolutely cannot substitute for good visuals in helping the reader to get to know the product. Figure 5.1 shows a typical use of a graphic for parts identification.

Even if your product seems to your eyes so simple that no parts diagram is needed, put one in anyway. It's pretty frustrating for a first-time user to be confronted with

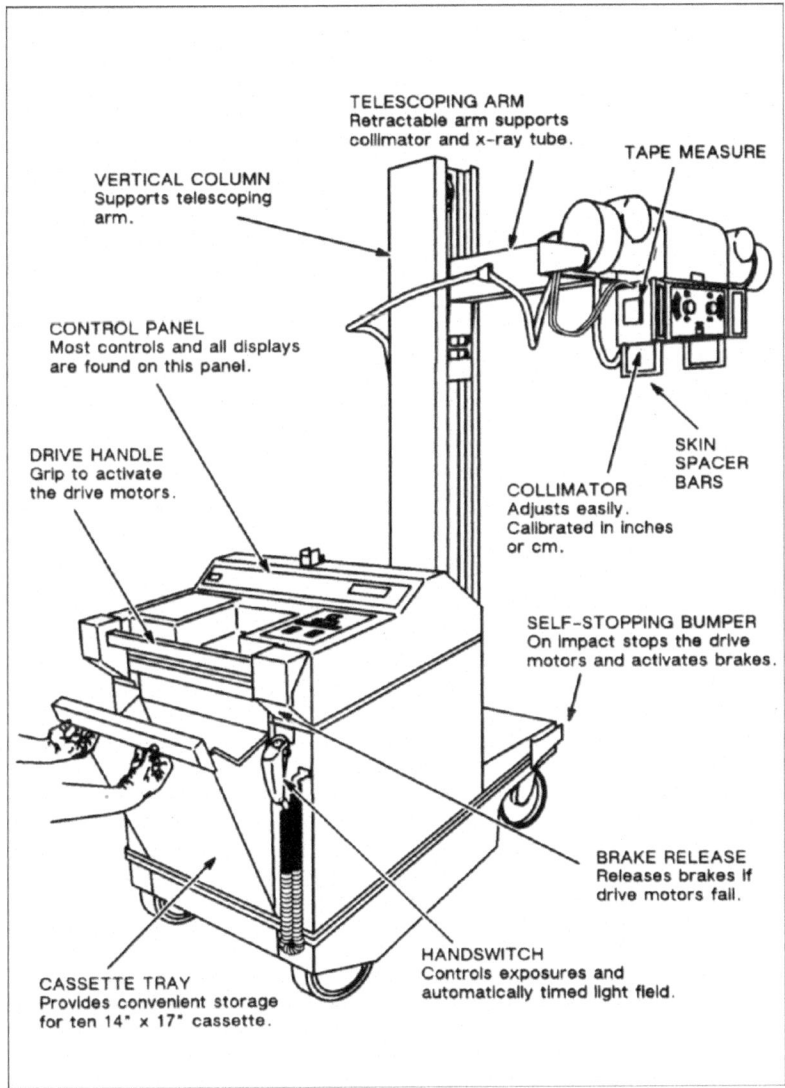

FIGURE 5.1   Typical use of a graphic to identify parts.

Source: From *Direction 17291 Revision D, AMX 4 Operation*, General Electric Medical Systems, Milwaukee, WI, 1989, p. 4. With permission.

unfamiliar part names while trying to figure out how a product works. Remember that because you are familiar with the product, you may take for granted information that is not intuitively obvious to someone else.

Some manuals include a section that is a parts catalog—intended for use in ordering replacement parts. Sometimes the parts catalog is a separate publication. For relatively simple products, the diagram in the parts catalog section may suffice, but generally it's better to have one or more visuals specifically designed to familiarize the new user with the product's major parts and controls. The diagram(s) in the parts catalog may be far more detailed than needed for operation, and may serve to confuse more than to clarify.

Explaining Spatial Relationships

As with parts identification, it is far easier to show where two parts of a product are in relation to each other than to describe it in words. Graphics can become especially important when a description depends on the product's being oriented in space in a specific way. What constitutes the "left" side, for example, changes with where one is standing in relation to the product. A good graphic can save many words of explanation.

Even such mundane matters as which way to turn a knob or to lift and tilt a cover to remove it can often be shown more effectively in a graphic. Consider an adjustment ring that must be rotated. If you try to use words alone, you may find yourself writing, "turn the adjustment ring to the right," or "turn the adjustment ring clockwise." The problem is that with anything round, if you turn the top of it to the right, the bottom goes to the left. And which direction is "clockwise" depends on which side of the adjustment ring you are looking at (as anyone who has ever struggled to disconnect two hoses coupled together can attest). How much easier for the writer and the reader if you were to include a drawing or photo with a curved arrow showing the correct way to turn the ring! See Figure 5.2 for an example of effective use of directional arrows in a visual.

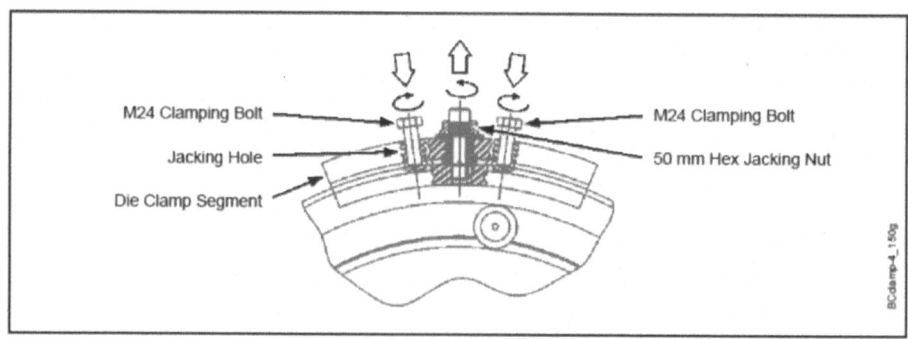

FIGURE 5.2 Use of arrows to show directional information.

Source: From *Installation, Operation & Maintenance Manual, 7900 Series Pellet Mills*, p. 5–32. © 2015 by CPM Acquisitions Corp. Used with permission.

Do You Have a Lot of Numbers to Convey?

Manuals typically contain a good deal of quantitative information. This information can range from gear ratios to baud rates, from load limits to vacuum pressures. Generally, quantitative information is much more effectively conveyed in the form of a table, chart, or graph than in words. This is especially true if the reader needs to be able to compare the quantities or otherwise relate the information. Numbers buried in paragraphs of prose are easy to miss and hard to relate to one another. Just as putting instructions in a series of numbered steps makes it easy for the reader to follow, putting information that the reader must relate (such as operating conditions or frequency of lubrication) in a table or chart makes it easy to see the connections.

The connections become obvious because charts and tables, like drawings and photos, can be seen as a whole. Closer inspection may be needed to see all the details, but the entire graphic can be seen all at once. Prose, on the other hand, is linear: you can skim it, but cannot see it all at once. The information is presented sequentially. As a consequence, paragraphs full of numbers become very confusing, forcing the reader to try to remember all the numbers until the relationships have been explained. It is especially difficult if different kinds of numerical information are presented—percentages and costs, for example—because typically like units are not next to each other for easy comparison, but rather in separate sentences. Such information is far easier to grasp when presented in a table or graph.

Explaining Procedures: Show and Tell

A standard classroom exercise to illustrate the importance of clear instructions is to have one student give oral instructions to perform a procedure such as tying a necktie or making a peanut butter sandwich. The exercise is set up so that the student giving the instructions cannot see the student carrying them out. The rest of the class can, however, and the result is always entertaining. It points up the fact that sometimes it's far easier to show someone how to do something than it is to tell how to do it.

In manuals, the photographs and drawings do the "showing." With a graphic, a writer can show a user where to place a tool, how to test a belt for proper tension, or how to identify a hose that's ready for replacement. A series of pictures can lay out a sequence of actions, almost as if the user had an expert-in-residence to demonstrate the procedure in person. We know that people would almost always ask another person to show them how to do a procedure rather than study a manual; well-conceived illustrations can be the next best thing.

Will the Manual Be Translated?

Graphical elements take on even more importance when the manual is written for an international market. Selling products overseas (or even in Canada) commonly means producing the manual in more than one language. Well-designed and well-reproduced graphics can make translation more cost-effective for international manuals. The reason is that text written in English tends to expand when translated into any other language. English has a very rich vocabulary, with many near-synonyms, differing only in nuance or connotation. Thus, one can be very precise (by choosing just the right word) and, at the same time, very concise (since "just the right word" exists). Other languages, with a smaller lexicon of words to choose from, require more words to attain the same precision. Thus, producing foreign-language text usually means higher printing costs on top of translation costs. "Pictures" do not need to be translated like text (although the callouts and captions do!), so designing graphics to carry a significant portion of the information makes good economic sense.

Note, however, that graphics do have cultural components that must be taken into account. As noted in Chapter 1, some visual depictions, such as images of a hand or pointing finger, can be offensive in some cultures. But you must also be certain that any symbol—that is, non-representational "art"—that you use works well in other cultural contexts. Not all cultural contexts are defined geographically. As technology continues to advance at a furious pace, choosing appropriate symbols across age-group "cultures" can be challenging. How would you graphically represent a computer? As a desktop model with separate monitor and keyboard? As a laptop? Or as a tablet? Choice of symbols becomes especially important when designing warning labels, as Chapter 9 discusses.

WHAT KIND OF GRAPHICS SHOULD I USE?

Just as a writer can usually think of two or three different ways to say the same thing, a visual designer can imagine two or three different ways to show the same thing. How do you choose whether to use a photograph or line drawing or a table or chart? Each type of graphic has advantages and drawbacks. The trick is to choose the right kind of visual for the purpose you have in mind.

PHOTOGRAPHS

Photographs are among the most common kinds of graphics used in manuals. Their chief advantage is that they are easily understood by the most technically naïve user. A photograph of a product or part looks like that product or part. The user needs no special training to interpret a photo. Additionally, photos make it easy to show a part in context, surrounded by other parts of the product. This can aid a new user in identifying unfamiliar parts.

The realism of photos carries with it two drawbacks. First, photos may easily be cluttered and fail to isolate the intended part from the context. This problem can often be solved by simply putting a contrasting outline around the part or pointing it out with an arrow. Or you can airbrush away the background distractions. Second, photos do not lend themselves to hidden views. If you can actually slice a part in half, you could photograph the resulting "cutaway" view, but drawings are more commonly used to show hidden views.

In the past, photographs had another disadvantage: they were expensive to reproduce, because they required an extra step before printing. Because photos contain "continuous tones," that is, shades of gray, a process is needed to convert those grays to a form that can be printed with black ink on white paper. Photos used to be projected through a halftone screen, which converted the image to a pattern of dots of variable size and spacing. At normal reading distance, our eyes would see shades of gray rather than individual dots. If a photo needed any retouching, such as airbrushing, that process added to the cost of production as well.

As with so much else in the world of technical publications, the digital era has changed all of this. While halftone images are still being produced, they are almost always generated by software producing digital images rather than screens and film emulsions. High-resolution digital cameras have now become quite affordable. Photo-editing software enables anyone to optimize exposure and retouch—even "airbrush" photographs. As a result, it is now usually cheaper to use photographs in a manual rather than drawings, which require a graphic artist to produce high-quality images (which, of course, will probably be drawn with a computer mouse rather than a technical pen!).

If you use photographs, you must make sure that they reproduce well. Digital photos shot at too low a resolution will look grainy, like a film image shot through a too-coarse halftone screen or with film that is too "fast." When an image is digitized, it is divided into tiny segments, called pixels, rather like the squares on a checkerboard. Resolution is usually expressed in pixels wide by pixels high—the checkerboard would have a resolution of 8×8. The more pixels, the finer-grained the picture. At the same time, the more pixels, the bigger the graphics file taking up space on the hard drive or in the cloud. Do not reduce resolution to save digital space—photos in a manual should be of sufficient resolution that the user can see needed detail. Nothing is more frustrating to a reader than being unable to see what a callout arrow is pointing at because the picture is too fuzzy to tell. Poor resolution can make even properly focused photos appear fuzzy.

Resolution is only half of the equation in manual photos. The other half is printing. Printers are usually described in "dots per inch" (dpi) or "pixels per inch." If you have a 200 dpi printer, a relatively low-resolution photo, say 640×480 pixels, can be printed no larger than about $3\frac{1}{2} \times 2\frac{1}{2}$ inches or else the individual pixels will start to be visible. Be sure that your photographs and printing capabilities match. An often-overlooked attribute of manuals that can affect the appearance of photos is the paper they are printed on. As noted in the last chapter, paper that is too porous can allow ink to spread and "bleed," making even the best photograph look fuzzy and out of focus.

Whether you create your own photographs or work with a professional, make sure that the photos have adequate contrast and proper exposure. To some degree, problems in these areas can be fixed in the lab or on the computer, but they are best addressed initially when the shot is taken. Arrange for good lighting. Beware of shadows that obscure the very detail you are trying to show, or glaring highlights that ruin the light balance and make everything else too dark. Walk around the product to find the best angle to shoot from. Think about whether a single photo will do, or whether you need a wide view to show something in context and a close-up for detail. Even if you work with a professional photographer, your job is to direct the professional to get the images you need. As suggested throughout this book, put yourself in the user's shoes. Plan your photos to serve the user's needs.

DRAWINGS

Unlike photos, which are called "continuous-tone" art (containing shades of gray), drawings are line art, meaning that they can easily be printed as a series of solid-color (usually black) ink lines on a page. Drawings do not require special treatment—camera-ready line art is no different from camera-ready text. Both can be turned directly into a printing plate with no special treatment. Ease of reproduction is only one advantage.

Drawings are widely used in manuals for a number of reasons. One big reason is that you can design a drawing to show exactly what you want, whether that view is normally visible or not (see Figure 5.2 and Figure 5.3). You can clear away extraneous clutter that is there in real life. You can show a cutaway view to point out hidden parts. You can create an assembly diagram, also called an "exploded view," to show how parts go together. Drawings can be created relatively easily on a computer and stored electronically.

So why doesn't every manual use only drawings? The answer, of course, is that they have drawbacks as well as virtues. One major drawback is that they are expensive to produce, since they require a graphic artist to draw them. While software is

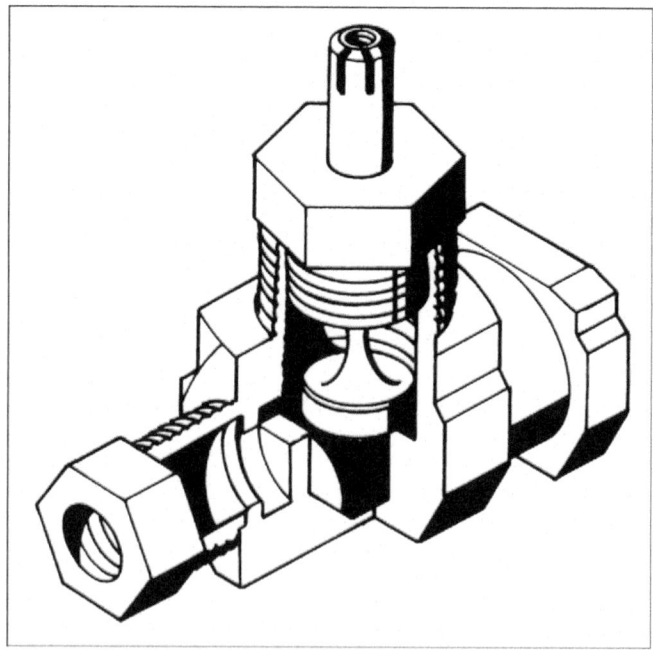

FIGURE 5.3   Cutaway view of a valve. A cutaway view allows the reader to see the construction of an object by presenting it as if a portion were literally cut away, revealing the layers of composition. The cutaway view is particularly useful to show the interior of something that is not normally disassembled.

Source: Drawing by Teresa Sprecher.

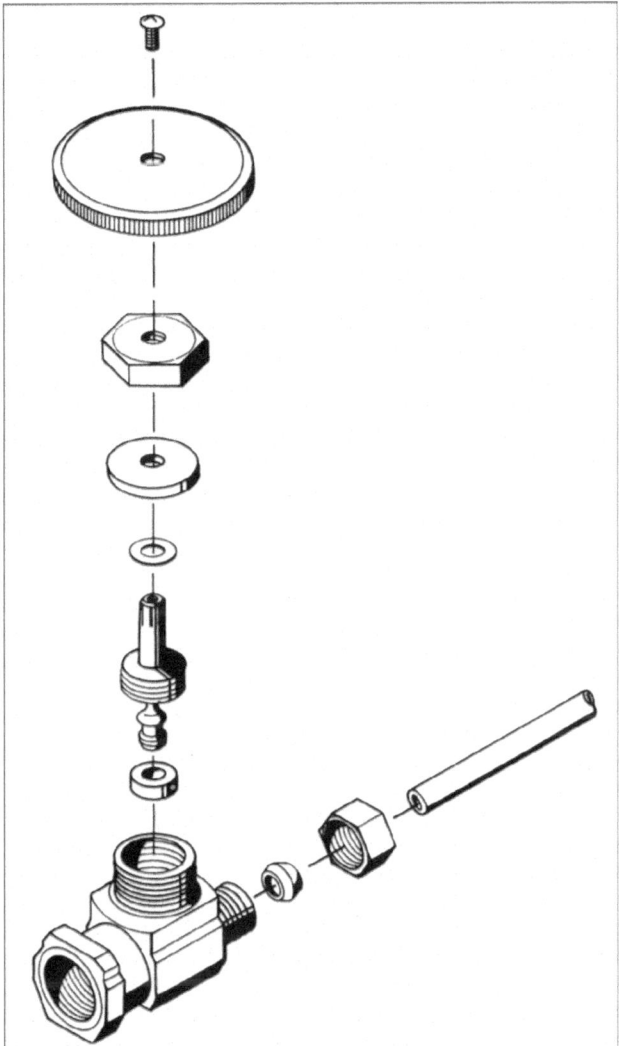

FIGURE 5.4   Assembly diagram of a valve. An assembly diagram uses an "exploded" view to let the reader see how the parts of an assembly fit together. It is used most often in conjunction with instructions for assembly or disassembly procedures.

Source: Drawing by Teresa Sprecher.

available that anyone with a mouse and monitor can theoretically use, a professional will design and render a much better graphic than an amateur. The professional will also do it faster. Do not try to save money by having the writer (whose expertise is words) try to create something usable with a computer drawing program! He or she will spend hours and the result is likely to be disappointing. This is not to say that the writer/artist with skills in both areas does not exist—just that such gems are rare.

To solve the cost problem, manual designers are sometimes tempted to use design drawings for the product as manual illustrations. After all, these already exist and they were drawn or produced on a computer-aided design (CAD) system by professionals. Unfortunately, design drawings are usually not suitable for manual illustrations. Any drawing is an abstraction and requires some visual sophistication to interpret. A photograph looks like the product—most people can understand a photo. Drawings, on the other hand, typically look quite different from the product. Many of the visual cues that we rely on to give an image its "three-dimensionality," such as shadows, foreshortening, and texture, may be absent from a drawing. They are almost certainly going to be absent from a design drawing. See Figure 5.5 for an example of a design or engineering drawing.

In addition, design drawings are generally quite large, as big as 18 × 24 inches or even larger. To fit onto a typical manual page, they need to be reduced drastically. When drawings are reduced, detail is lost, they become difficult to read, and the accompanying lettering becomes impossible to read. If you ever do use a design drawing as an illustration in a manual (an electrical schematic in a service manual, for example), consider redoing any callouts or other lettering after the reduction has been made.

The other option, used commonly in industrial maintenance manuals, is to keep the drawing its original size, and make it a foldout. This option should be used sparingly, and certainly not ordinarily in an operator manual. Odd-size pages are difficult to manage and easily get torn.

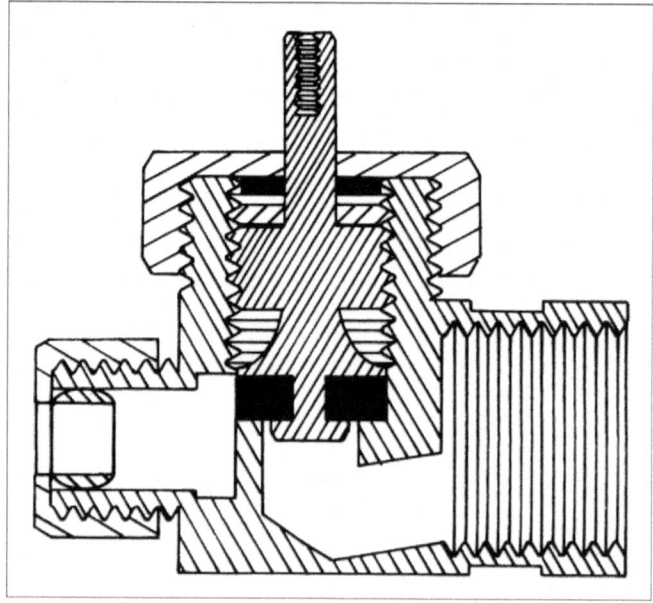

FIGURE 5.5   Design drawing of a valve. This design drawing, also called an engineering drawing, of the valve is much more abstract and requires special training to interpret.

Source: Drawing by Teresa Sprecher.

For many applications, the best choice of line drawings is a simple perspective drawing, as shown in Figure 5.6. This type of drawing adds back some of those visual cues that make it easy to interpret, but it has the added advantage of allowing you to focus on the important part and reduce background clutter.

TABLES

Tables allow you to show a great deal of quantitative information in a very compact form. For example, Figure 5.7 shows thread and torque specifications for eight sizes of

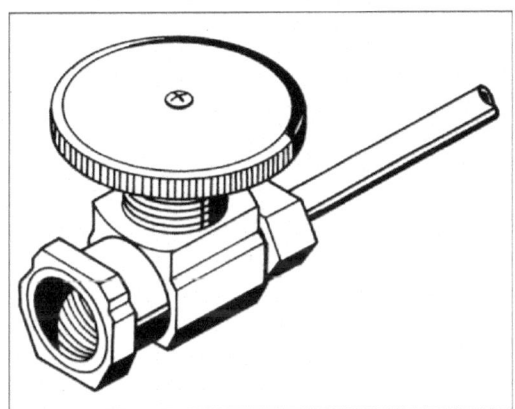

FIGURE 5.6   Perspective line drawing of a valve. This drawing is an abstraction—it is not *really* the valve—but it is quite recognizable to an untrained eye.

Source: Drawing by Teresa Sprecher.

Table 1. CPM Metric Fastener Standards: Hex Head Cap Screws

Metric Size	Thread MM Per Pitch	Grade 8.8 Tightening Torque Ft-Lb	(M-Kg)	Wrench Size (MM)	Grade 10.9 Tightening Torque Ft-Lb	(M-Kg)	Wrench Size (MM)	Similar Inch Size
M6 X 1	1.0	7	(1)	10	10	(1.4)	10	1/4
M8 X 1.25	1.25	8	(2.5)	13	25	(3.5)	13	5/16"
M10 X 1.2	1.5	32	(5)	17	47	(6)	17	3/8"
M12 X 1.75	1.75	58	(8)	19	83	(12)	19	1/2"
M16 X 2.0	2.0	144	(20)	24	196	(27)	24	5/8"
M20 X 2.5	2.5	260	(36)	30	366	(51)	30	3/4"
M22 X 2.5	2.5	368	(51)	32	520	(72)	32	7/8"
M24 X 3	3.0	470	(65)	36	664	(92)	36	1"

Die Clamp Bolts may require special tightening torque. See the Die Installation procedure elsewhere in this manual.

FIGURE 5.7   Table showing data in a compact form. The use of a table permits a great deal of information to be packed into a small space.

Source: From *Installation, Operation & Maintenance Manual, 7900 Series Pellet Mills, Introduction to Metric Fasteners*, p. 2. © 2015 by CPM Acquisitions Corp. Used with permission.

screws in two grades—all in the space about the size of a 3" × 5" index card. Imagine the confusion that would result if you tried to present all that information in paragraph form!

Useful as tables are, they must be designed well if they are to work well. The following list presents some guidelines for constructing tables:

- Arrange the headings and data in a rational order.
- Display items vertically that you want your reader to compare directly. Most people find it much more difficult to compare information that is arranged horizontally (see Figure 5.8.)
- Include units of measurements in headings, rather than in each cell of the table—it will reduce clutter.
- Align columns of numbers on the decimal point.
- Use lines to divide columns and rows only if confusion is likely without them. Lines add clutter and tend to draw the eye along them rather than across them. Try to use white space as a divider instead. Note that you can eliminate gridlines between columns only when the space between the columns itself makes a more or less "straight line." If you arrange direct comparisons vertically, you will more likely have even columns, since the elements will be similar. See Figure 5.9 for examples of tables with and without dividing lines.

PARDEE SPINNING RODS

MODEL	LENGTH (FEET)	LINE WT (LB-TEST)	LURE WT (OZ)	ACTION	PRICE
UL-500	4.5	2–4	1/16–1/8	LIGHT	$32.50
UL-600	5	2–4	1/16–1/8	MEDIUM	$38.00
WB-650	6	6–8	1/8–1/4	MEDIUM	$43.00
WB-800	6.5	6–8	1/8–1/4	STIFF	$47.50
BN-850	6	8–12	1/4–3/8	MEDIUM	$52.75
MS-900	6	12–16	1/4–1/2	MEDIUM	$63.00
MS-1000	6.5	12–20	3/8–5/8	STIFF	$72.00

(VERSION A)

PARDEE SPINNING RODS

	UL-500	UL-600	WB-650	WB-800	BN-850	MS-900	MS-1000
LENGTH (FEET)	4.5	5	6	6.5	6	6	6.5
LINE WT (LB-TEST)	2–4	2–4	6–8	6–8	8–12	12–16	12–20
LURE WT (OZ)	1/16–1/8	1/16–1/8	1/8–1/4	1/8–1/4	1/4–3/8	1/4–1/2	3/8–5/8
ACTION	LIGHT	MEDIUM	MEDIUM	STIFF	MEDIUM	MEDIUM	STIFF
PRICE	$32.50	$38.00	$43.00	$47.50	$52.75	$63.00	$72.00

(VERSION B)

FIGURE 5.8 Two versions of a table showing the importance of vertical comparison.

LINE CAPACITY OF SPOOL (IN YARDS)		
Lb. Test	Large	Small
2	---	---
4	400	300
6	350	200
8	300	150
10	250	100
12	200	---
15	150	---
18	100	---

LINE CAPACITY OF SPOOL (IN YARDS)		
Lb. Test	Large	Small
2	---	---
4	400	300
6	350	200
8	300	150
10	250	100
12	200	---
15	150	---
18	100	---

PARTS LIST (PARTIAL)			
Part Name	Order No.	Part No.	Price
Axle	305	81-214	$2.25
Baffle Plate	6342-R	56	1.20
Click Spring	11	322-D	0.30
Drive Gear	452-T	9120	4.50
Pivot	6783-DE	3	0.75
Rotating Head	66-2	81-615-A	11.45
Spool	43598-OS2	27	3.75
Transfer Gear	452 (453)	16 (17)	0.75 (0.85)
Trip Lever	340972	81-005	1.25

PARTS LIST (PARTIAL)			
Part Name	Order No.	Part No.	Price
Axle	305	81-214	$2.25
Baffle Plate	6342-R	56	1.20
Click Spring	11	322-D	0.30
Drive Gear	452-T	9120	4.50
Pivot	6783-DE	3	0.75
Rotating Head	66-2	81-615-A	11.45
Spool	43598-OS2	27	3.75
Transfer Gear	452 (453)	16 (17)	0.75 (0.85)
Trip Lever	340972	81-005	1.25

FIGURE 5.9   Two versions of two tables—with and without lines dividing the columns. Note that the use of lines is essential when the white space between columns is irregular—but is a distraction when the white space forms a sufficient visual barrier.

Tables are essential for presenting quantitative information, but that is not their only use. Other kinds of information can also be presented in table form. For example, Figure 5.10 shows a typical troubleshooting section set up as a table. Another example is Figure 5.11, which shows a comparison between two printer models in tabular form. Note that the entries are all words rather than numbers. Still, it makes the comparison much clearer than if the writer had tried to write it all out in paragraph form.

CHARTS AND GRAPHS

Charts and graphs are similar in that they are considerably more abstract than photographs or drawings, but they are not exactly the same. Charts show various kinds of information in symbolic form. In other words, charts use geometric shapes (circles, bars, diamonds, etc.) to stand for something rather than presenting a realistic image of the product. For example, you might use a flowchart to portray a diagnostic procedure (see Figure 5.12). Whatever sort you decide to use, make sure it will be easy for your user to understand. Keep it focused, uncluttered, and make sure that there is sufficient white space so that your reader can easily find the needed information.

Graphs show numerical information in graphical form. For example, you could use a bar graph to show the appropriate temperature ranges for various lubricants (see Figure 5.13). Sometimes quantitative information is more effectively presented

PROBLEM	POSSIBLE CAUSE	POSSIBLE SOLUTION
No filling.	Filler nozzle is not primed.	Run about 4 tortillas through to prime.
	Filler metering cylinder is not primed.	Follow instructions under "Set-Up."
	Filler hopper interlock is not being made.	Make certain hopper is properly located & antirotation pin is fully engaged.
	Completion of stroke sensor has not been activated.	Use screw driver blade or wrench to push on pilot button.
	Air is off to machine or filler.	Check & turn on.
First fold & fill keep cycling.	Water droplets on electric eye lens.	Clean & dry lens.
	Residue buildup on 1st conveyor belt.	Clean off belt.
	Sensitivity set too high on 1st electric eye.	Adjust so eye only detects passing of tortillas.
Loosely folded burritos.	Tortillas are stale.	Use fresher tortillas. Best results are obtained when tortillas are only a few hours old.
	Back Stop on 2nd fold platen is too far back.	Adjust position to lessen surface area on platen.
	Too small amount of filling.	Adjust filler per pages 16 & 17.

FIGURE 5.10 Troubleshooting chart in traditional format. The traditional format works well when there is no need for a sequential approach to troubleshooting.

Source: From *Model 323–1 Tortilla Folder, Technical Service Manual*, Kartridg Pak Company, Davenport, IA, 1990, p. 34. With permission.

	Epson GL/2	HP LaserJet III
PCL mode	Does not exist	Exists as the initial mode
Paper eject	Supports PG, AF commands	Supported in PCL
Auto eject	SelecType setting	Not available
Reduced printing	SelecType setting	Available in PCL
Switch to PCL (ESC % # A)	Not supported	Supported
Reset (ESC E)	Ejects paper and then initializes	Ejects paper, switches to PCL, and then initializes
PJL, EJL, and ES	Supported	Supported
Advance Full Page (PG, AF)	Supported	Not Supported

FIGURE 5.11 Table used to make comparison clearer. This example, taken from a printer manual, compares two printer modes. Even though all the entries are words, the table makes the comparison much clearer than if the information were written in paragraph form.

Source: From *Reference Guide, Epson ActionLaser 1000/1500*, Epson America, Inc., Torrance, CA, 1992, p. B-45. With permission.

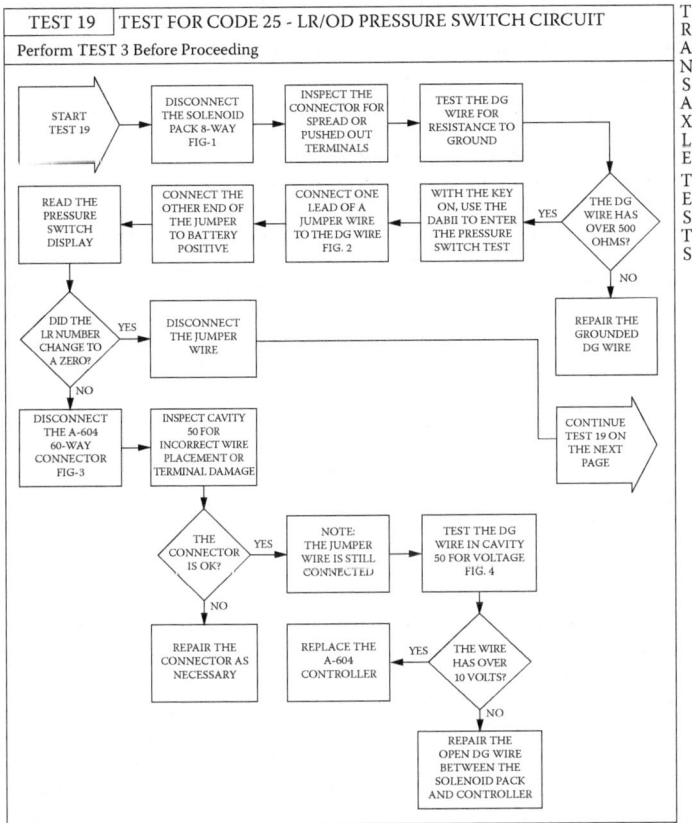

FIGURE 5.12 Diagnostic testing procedure shown as a flowchart. The flowchart format makes it very easy to see the sequence of tests.

Source: From *Powertrain Diagnostic Procedures A-604 Ultradrive Automatic Transaxle (1989), No. 81–699–9009*, Chrysler Motors Corporation, Center Line, MI, 1988, p. 45. With permission.

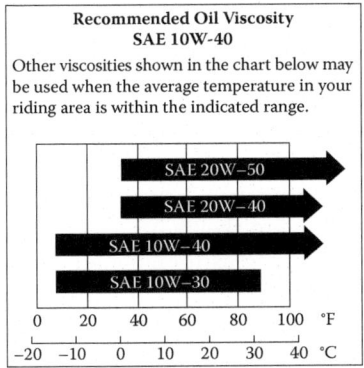

FIGURE 5.13 Bar chart used to show lubricants for different temperature ranges.

Source: From *Honda Owner's Manual 97 VT1100C2 Shadow 1100*, American Honda Motor Company, Torrance, CA, 1996, p. 70, © Copyright by American Honda Motor Co., Inc. With permission.

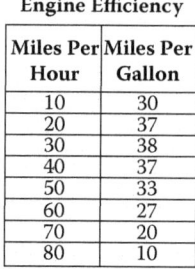

Engine Efficiency

Miles Per Hour	Miles Per Gallon
10	30
20	37
30	38
40	37
50	33
60	27
70	20
80	10

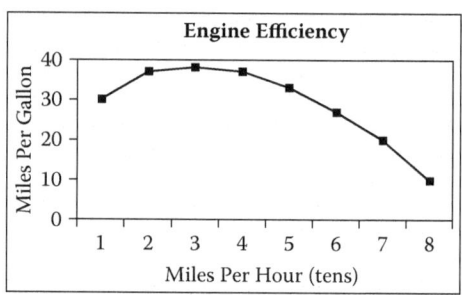

FIGURE 5.14   Numerical information shown in table and graph form. Notice how the relationship between the values is much easier to see when the data are presented as a line graph.

as a line graph than in a table. Consider, for example, the relationship between fuel consumption versus speed for a car engine. You could show this relationship by creating a table, in which you listed various speeds in the first column and miles per gallon in the second. But because the relationship is not strictly linear—because of gear ratios, the engine may consume more fuel at the lowest speed than at a moderate one—the relationship would not be easy to sort out. By contrast, if you plotted the same information as a line graph, with speed along the x-axis and fuel consumption on the y-axis, the relationship becomes immediately obvious. Look at Figure 5.14, which shows these two options.

MAKING GRAPHICS WORK

Once you have decided on the purpose of a graphic, you can begin to make the design choices that will ensure the graphic works—because you know what is important. The essence of good visual design may be summed up in the following fundamental rule: *make the important things stand out.* For example, you may have a choice between a photograph and a line drawing. As the previous section shows, the two media have different strengths. In one case, you may decide that it is most important to show the context of a part realistically, and choose a photograph. In another case, you may wish to show detail that would work better in a drawing. As far as time and budget permit, make these decisions for each illustration individually. Don't use an old photograph from another manual just because it is handy. In the long run, an illustration or table designed with a specific purpose in mind will better serve the user's needs and the company's interests. You may find it useful to sketch (or ask the graphic artist to sketch) different versions of a drawing or angles for a photograph that you have in mind, so that you can pick the one that works the best. After all, you wouldn't expect to be able to sit down at the computer and write final copy on the first try—graphics need "rough drafts," too.

Visual design is a major area of study, and certainly cannot be covered thoroughly in this book. The following general principles, however, apply to any visual presentation of information, regardless of type.

Size Matters: Make Them Big Enough to See

The user should easily be able to read and interpret a graphic at normal reading distance. Remember that a manual may be used in difficult conditions. Depending on the product, the user may be reading the manual in a basement or on a factory floor or in a dimly lit barn. Even if the lighting is good, it is crucial that the user be able to identify parts. If the graphic is too small (or reduced too much to fit the page), a machine screw may be indistinguishable from a large bolt.

The ability to digitally reduce and enlarge graphics is a wonderful thing, but it requires care in its use. In a photograph, excessive reduction will obliterate important detail. In a drawing, too much reduction can cause closely spaced lines to merge, and lettering to fill in, so that an e looks just like an o. Be particularly careful with large assembly drawings. Typically, a line weight that works well in the original becomes much too fine when it is reduced to page size. If the original drawing was done on a CAD workstation, you can go in and change line weights to suit the smaller size better. If the drawing was done by hand, it would be better to redesign it specifically for the manual.

Often, you can solve the problem of loss of detail from reduction by adding a close-up or inset. For example, Figure 5.15 shows the use of a close-up to show detail.

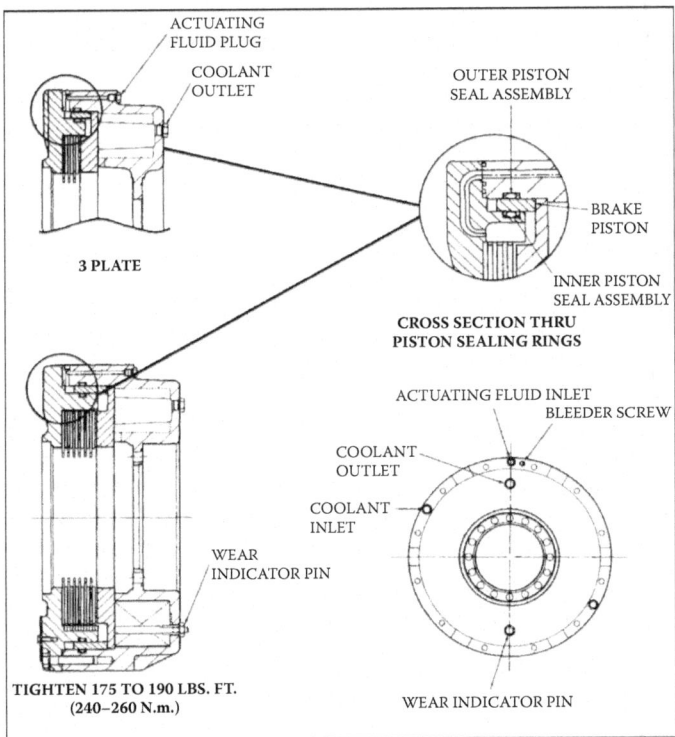

FIGURE 5.15   Use of a close-up to show detail. Note how the use of a close-up permits much more detail to be shown than would be possible otherwise.

Source: From *LCB 13150/16150 Posi-Stop Liquid Cooled Brake Maintenance and Service Manual Supplement*, Clark Components International, Statesville, NC, 1989, p. 4. With permission.

Be sure that if you use this technique, the user can tell where the part shown in the close-up view is located on the product. A nice close-up view of the power-steering fluid filler cap is not much use to the car owner if he can't find it when he looks under the hood!

Less Is (Still) More

Just as is the case with text in a manual, you must focus your graphics on the need-to-know. Without question, the most common fault of illustrations in manuals is that they are cluttered. A user will find it difficult to focus on (or even figure out) what is important in a graphic overloaded with information. Illustrations and charts must be edited just like prose: figure out what the purpose of the graphic is and then include only what is necessary to fulfill that purpose. You should not include everything you know in a graphic—any more than you would in a paragraph. Simplify the visual presentation so that only essential items are included in detail and non-essential items are either absent or merely suggested. If, despite your efforts to keep it simple, you still seem to have a complex illustration, consider splitting it in two—you may have more than one visual focal point. Break the presentation down by systems or show one overview and one or more close-ups. This is an especially important technique to use in parts catalogs, which too often show the entire product in a single incomprehensible assembly drawing.

This process of simplification is terribly difficult—the temptation is great to include more than you should—but you will find that knowing the purpose of the visual will help enormously. If you have a clear idea of what you want the drawing, photo, or chart to accomplish, you can use that idea as a filter to screen out peripheral information, much as you would use a topic sentence to guide your choice of what to include in a paragraph. For example, the block diagram shown in Figure 5.16 is perfectly adequate for explaining the theory of operation of the transaxle in this service training manual. The full schematic (Figure 5.17) is included at the end for reference.

What Am I Looking At? Making Graphics Clear

Be sure that each drawing, photo, table, and chart has a title that tells what it shows as well as a figure number or table number. For example, "Figure 3, location of idle adjustment screw" is much better than just "Figure 3." Pay special attention to how you "call out" items in the illustration. Here, as always, there are trade-offs to be made. For example, compare Figure 5.18 and Figure 5.19. They are very similar illustrations showing the controls for a motorcycle. The first one, from a Honda manual, uses part names and arrows to call out the parts; the second, from a Suzuki manual, uses numbers and lines (with no arrowheads) to call out the parts. The Honda illustration, because it uses part names, saves the user a step in getting the information. The user does not need to go to the legend to identify a part. On the other hand, the Suzuki illustration appears less cluttered and more focused on the drawing. Furthermore, it would require no change to the illustration itself for use in a translated manual—only the legend would need to be changed.

Which is better? The answer depends on your analysis of the audience and purpose for the graphic. Is it more important for the user to be able to look right at the

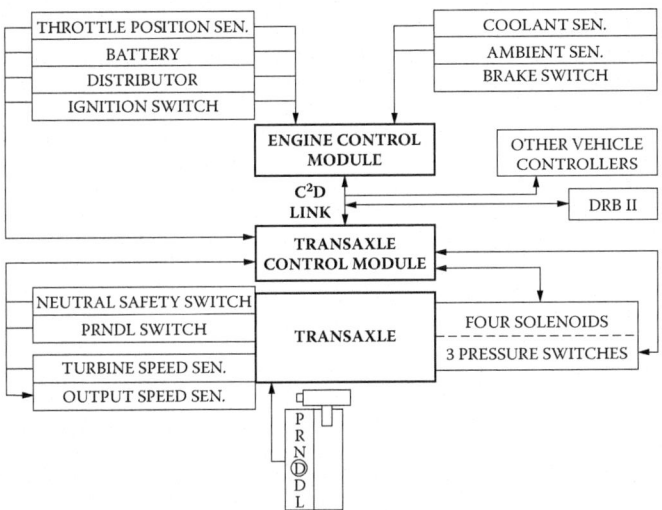

FIGURE 5.16 Block diagram to indicate electrical components. At this point in the manual, a block diagram is sufficient. Using a block diagram permits the reader to concentrate on the important parts and avoid being distracted by detail.

Source: From *Powertrain Diagnostic Procedures A-604 Ultradrive Automatic Transaxle (1989),* *No. 81–699–9009,* Chrysler Motors Corporation, Center Line, MI, 1988, p. 19. With permission.

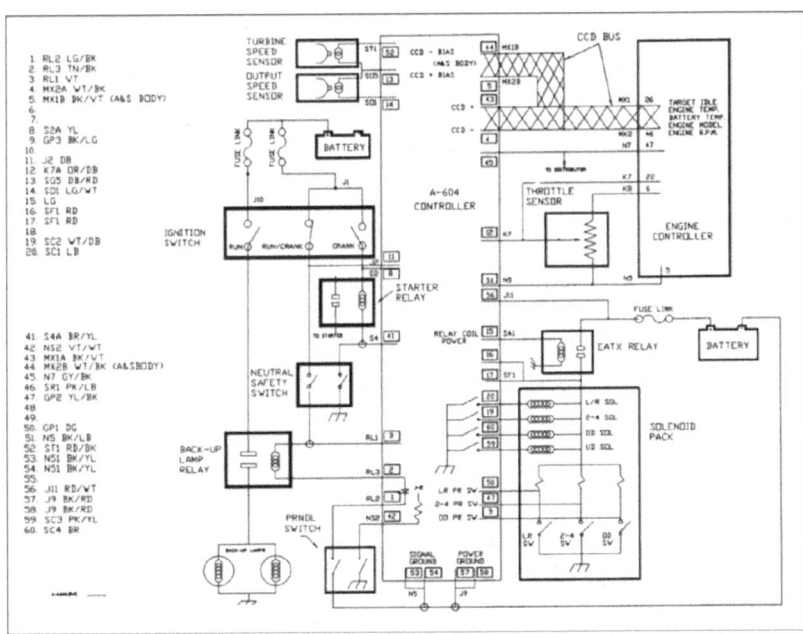

FIGURE 5.17 Full schematic to aid in servicing an electrical system. Here the schematic is necessary for completeness.

Source: From *Powertrain Diagnostic Procedures A-604 Ultradrive Automatic Transaxle (1989),* *No. 81–699–9009,* Chrysler Motors Corporation, Center Line, MI, 1988, p. 118. With permission.

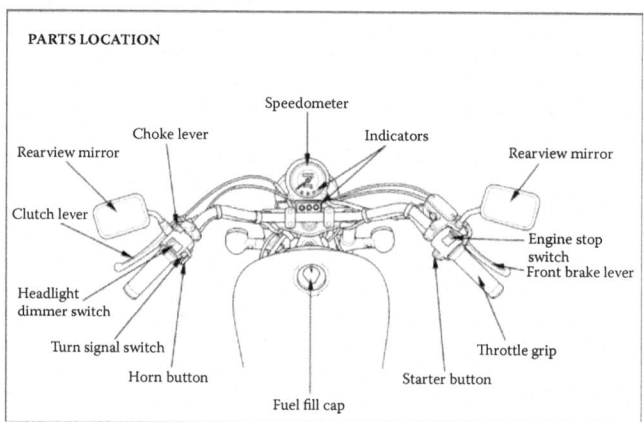

FIGURE 5.18   Illustration using words to call out part names. Using words and arrows to call out parts makes it quicker for the user to find the part name, but limits the amount of information that can be shown in one drawing and requires an extra step for translation.

Source: From *Honda Owner's Manual 97VT1100C2 Shadow 1100*, American Honda Motor Company, Torrance, CA, 1996, p. 11, © Copyright by American Honda Motor Co., Inc. With permission.

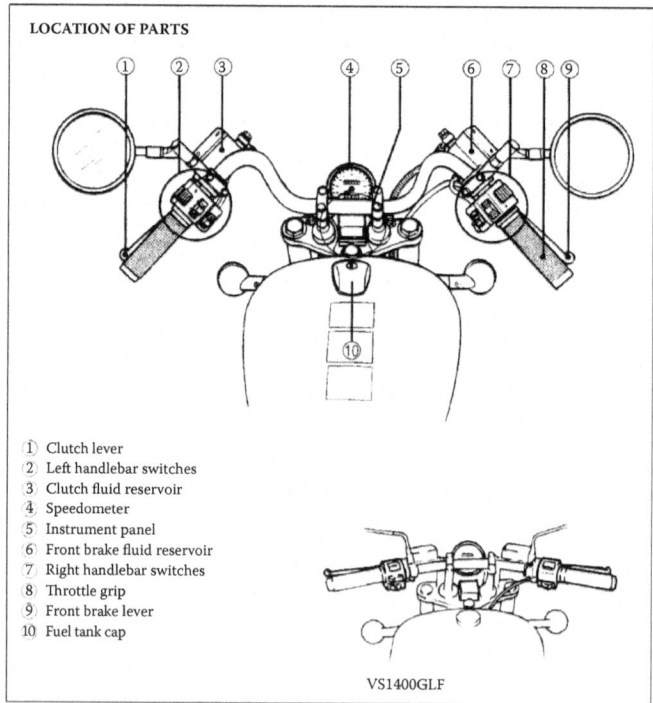

FIGURE 5.19   Illustration using numbers to call out part names. Using numbers to call out parts requires the reader to go through an extra step, but allows more information to be included in a single illustration, and simplifies translation.

Source: From *owner's Manual, Suzuki VS1400GL*, American Suzuki Motor Company, Brea, CA, 1997, p. 10. With permission.

drawing and identify parts without having to look away at a legend, or is it more important for the drawing to appear uncluttered? An experienced motorcycle rider might not be bothered by clutter, since he or she is already familiar with the controls that are usually around the handlebars, and would be looking primarily to find any changes from what he or she was used to. On the other hand, a novice might be holding the book while finding parts on the motorcycle. In that case, having to look at the legend would probably not be a problem, since the user would be looking back and forth between the product and the manual anyway. How important is it that the manual be easily translated? If marketing plans call for translation into several languages, the numerical callouts will save money.

Whichever way you choose, make it as easy for the reader as possible. Be certain that it is clear what part the arrows point to. If you use numbers or letters rather than words to call out parts, arrange them in a rational order: try to put them in sequence. Usually, you will have to choose whether the callouts are in sequence or the parts in the legend are in some order (such as alphabetical). Make your decision on how the graphic will be used. If the user will be looking at the illustration first to locate the part, as would be likely in a parts catalog or in a procedure description, make the numbers on the illustration sequential. On the other hand, if the user would first be looking up a name in the legend, and then locating it in an illustration, make the entries in the legend sequential (alphabetically or by part number). In either case, keep the legend close to the illustration—on the same page if possible, on a facing page if not. Avoid making the reader turn the page to look at the legend.

Use the principles of good visual design to help focus the reader's attention and to avoid confusion:

- The eye moves along lines, not across them. Use line direction to lead your reader's eye to the central focus of the graphic.
- Bigger or more detailed objects will seem more important than smaller or less detailed ones.
- Similar shapes (in a block diagram, for example) will suggest similar function.
- Provide plenty of white space around and within a graphic. White space can help the user organize visual information, reduce clutter, and serve as a *de facto* "frame" around a graphic. Sometimes, all a confusing or cluttered illustration needs is a bit of white space to become more inviting to use.

INTEGRATING GRAPHICS AND TEXT

Writers, because they are comfortable with the world of words, may find planning and developing the text of a manual to be a relatively straightforward task, but be somewhat baffled by how to manage the graphical elements. Too often, the graphics become an afterthought, added in as the page layouts permit. As a result, they may be disconnected from the information in the text, sized too small to see easily, or too infrequent to be of much help. For graphics to work well in a manual, they must be part of the manual design from the beginning.

BEGIN AT THE BEGINNING: STORYBOARDS AND OUTLINES

The most effective manuals are planned from the first as an information package containing both text and graphics. As you start to block out the initial outline for a manual, be thinking about what graphical elements you would like to include. Any time you find yourself thinking "in pictures," plan to give the reader a picture as well.

A technique borrowed from filmmaking can be very useful in helping to integrate text and graphics in the early stages of designing a manual. Filmmakers prepare one or more "storyboards" for each scene of a film. These storyboards contain both visual and textual information. The technique translates easily to manual design: think of it as a step between the outline and a first draft. It works particularly well if you are planning a modular organization, as discussed in Chapter 3.

To make a storyboard, choose a small section of the manual—no more than a page or two in the finished manual—and prepare a plan for that section. The plan includes a prose summary of the material to be covered in that section and a sketch or description of an accompanying graphic. You can then lay out these storyboards in order, and see how the manual will develop. It is easy at this stage to rearrange sections if another order makes more sense. Storyboarding works best if both writer and artist can be involved at the earliest stage of the manual, helping to ensure that text and illustrations will balance and support each other.

WHAT ARE MY GRAPHICS FOR?

It's a good idea to jot down the purpose as well as the subject for both text and graphics. The difference between the two is subtle, but real. The subject tells what the graphic is *about*; the purpose is what the graphic is *for*. Regardless of the specific type of graphic (e.g., photograph, table, etc.), all graphics have one of three purposes:

- To complement text
- To supplement text
- To substitute for text

Graphics That Complement Text

Graphics that complement text are those that are necessary for the reader to understand what the text is saying. Figure 5.20 shows a page from a manual for a DR Brush Mower. Note how the instructions are accompanied by several photographs that direct the user how to carry out the instructions, by locating parts and showing procedures. The instructions would be unintelligible without the photos. Thus, in this situation, the graphic complements the text—which would certainly be incomplete without it.

This method of coupling one or more photos with specific instructions has been widely used and adapted in many industries. It has the advantage of making the manual appear very accessible, even for poor readers. It also reduces translation costs, since some of the burden of communication is on the photograph rather than the text. Because these text-and-graphic blocks can be developed and stored electronically as a single unit, they make modular organization easier than if the text and

Figure 29

Figure 30

Figure 31

To Replace the Drive Belt

Tools and Supplies Needed:

- Ratchet
- 1/2" Socket
- 5/8" Socket
- #2 Phillips Screwdriver
- Gloves

1. Remove the Brush Deck (See section "Removing the Brush Deck" on page 10) and tip the machine back on its handlebars to access the Clutch Connector under the machine (Figure 29).

2. Disconnect the Clutch Connector by lifting the locking tab and separating the two halves (Figure 30).

3. Locate the hole in Traction Drive Pulley (Figure 31 and 32) on the engine and insert the Phillips head screwdriver through the opening in the Frame and into the hole in the Pulley (Figure 31).

4. If the hole is not aligned with the Screwdriver, turn the Clutch Bolt with the 5/8" Socket until the screwdriver goes into the hole.

5. Rotate the Clutch Bolt (direction to loosen) until the screwdriver rests against the frame (this is to keep the engine shaft from rotating in the next step).

6. Remove the Clutch Bolt using a 5/8" Socket. The Clutch Bolt has standard, right hand threads (Use impact wrench if possible).

7. Remove Clutch from engine shaft.

8. Remove the nuts retaining the belt guide with a 1/2" socket (Figure 32).

9. Remove the Key from the Traction Drive Pulley and shift the transmission to Neutral **N**.

10. Rotate the Pulley as you pull the belt out of the pulley groove.

11. Remove the belt from the Transmission by rotating it 90° and sliding it between the Transmission Pulley and the Frame.

12. Reinstall the new belt by reversing the above procedures.

During reassembly make sure that:

- The Shaft key is installed in the engine shaft.
- The belt is on the inside of the belt guides (Figure 32).
- The clutch is located properly on the Anti-Rotation Bolt (Figure 33).
- You torque The Clutch Bolt to 50lb-fts (68N-m).

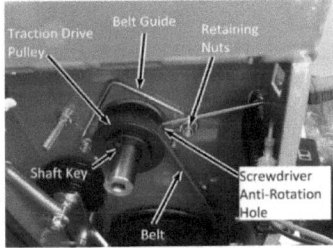

Figure 32

Figure 33

22   **DR**® FIELD and BRUSH MOWER

FIGURE 5.20   Use of graphics to complement text. Without the graphics to illustrate what the text is describing, the procedure would be difficult to follow correctly.

Source: From *DR© Field and Brush Mower Safety and Operating Instructions,* © 2017 by Country Home Products, Inc. All rights reserved. Used by permission.

graphics were developed separately. This format does have its limitations, however. Not all instructions can be broken down into small, simple steps. Sometimes complex explanations are necessary. And many kinds of instructions (for using software, for example) do not lend themselves well to accompanying photos.

Graphics That Supplement Text

Graphics that supplement text are those that serve primarily as illustrations: they add to the message presented in the text, but are not necessary for its understanding. A good example of such a graphic is shown in Figure 5.21. The photograph illustrates what the text describes, but the text still makes sense without it. While graphics that supplement text may not be required for the text to be understood, they certainly contribute to it. Such graphics are especially useful for first-time users.

Remember that no reader, regardless of how experienced, is as familiar with the product as you are. Having written the manual and (if you are an engineer-writer) possibly having designed it as well, it is easy to forget the questions and confusions of the user, especially those of a novice to the product. Try to look at your product

PLACEMENT OF HANDS DURING FEEDING

At the start of the cut, the left hand holds the work firmly against the infeed table and fence, while the right hand pushes the work toward the knives. After the cut is underway, the new surface rests firmly on the outfeed table as shown in Fig. 64. The left hand should then be moved to the work on the outfeed table, at the same time maintaining flat contact with the fence. The right hand presses the work forward, and before the right hand reaches the cutterhead, it should be moved to the work on the outfeed table.

CAUTION: NEVER PASS HANDS DIRECTLY OVER THE CUTTERHEAD.

JOINTING AN EDGE

This is the most common operation for the jointer. Set the guide fence square with the table. Depth of cut should be the minimum required to obtain a straight edge. Hold the best face of the piece firmly against the fence throughout the feed as shown in Fig. 65. Maximum depth of cut should not be more than 1/8" in one pass.

DO NOT PERFORM JOINTING OPERATIONS ON MATERIAL SHORTER THAN 10 INCHES, NARROWER THAN 3/4 INCH, OR LESS THAN 1/2 INCH THICK (REFER TO FIG. 66).

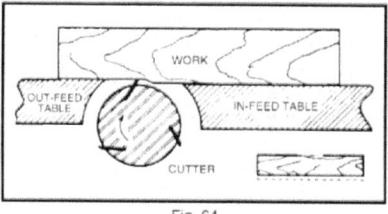

Fig. 64

Fig. 65

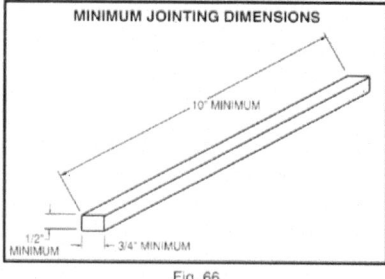

Fig. 66

FIGURE 5.21   Use of graphics to supplement text. In this example, the text is easily understood without the accompanying photographs, but they certainly serve to enhance it, particularly for a first-time user.

Source: From *Instruction Manual, 6" Deluxe Jointer (Model 37–190)*, Delta International Machinery Corp., Pittsburgh, PA, 1998, p. 22. With permission.

with fresh eyes—imagine it is your first look—and think about what photographs and drawings would help you to understand the instructions. This is simply another form of writing with the user in mind. In any case, don't skimp. Use as many graphics as you can to supplement the text. The short-term added cost will save money in the long run by reducing the need for technical assistance and service calls. As noted elsewhere, an easy-to-use manual will help build repeat business as well. Good illustrations are a big part of making a manual easy to use.

Graphics That Substitute for Text

Graphics that substitute for text are becoming more common. While few manufacturers design manuals that are all pictures and no words, many use graphics as a substitute for text in selected places. Figure 5.22, for example, uses an illustration to show how to hook up a scanner to a computer. The accompanying text describes USB connectors and how to configure the system to accept the new hardware, but does not instruct the user on the actual hookup.

Figure 5.23 shows a section of the assembly instructions for a gas grill. The instructions are presented almost exclusively in pictures rather than words.

When graphics are used to substitute for words, they must be very carefully designed. If the graphics are unclear in any way—too small to identify parts, poorly reproduced, too cluttered—then without text as a backup, the reader will be lost. It would be wise to test any such stand-alone graphics before using them in a manual. Be sure to pick a range of test subjects, and include some without prior experience with the product. Note also that users in Europe and the Pacific Rim countries may more easily use picture-only instructions than users in North America. The large number of different languages spoken in those areas has made it more common to rely on visual depictions for many communication tasks. International warning labels, for example, are commonly symbol-only, as discussed in Chapter 9.

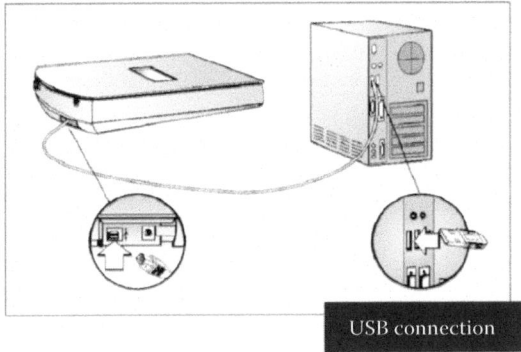

USB connection

FIGURE 5.22   Use of graphics to substitute for text. In this example, showing how to hook up a scanner to a computer, the illustration stands on its own. No words are necessary to understand the connection procedure.

Source: From *Installation Supplement, HP Scanjet Scanner*, Hewlett-Packard Company, Palo Alto, CA, 1998, p. 10. With permission.

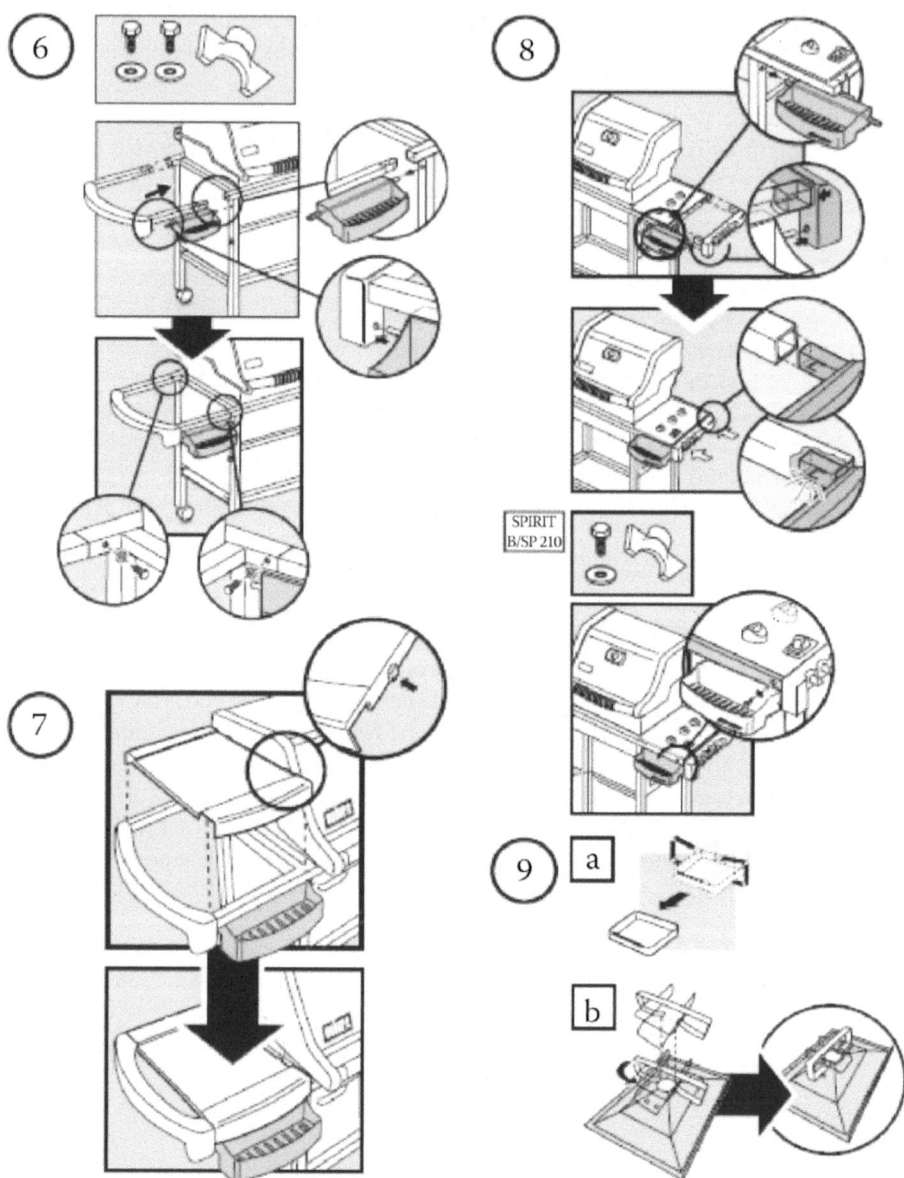

FIGURE 5.23 Example of primarily pictorial instructions. This excerpt from the assembly instructions for a gas grill relies almost exclusively on graphics to convey the information. Pictorial instructions reduce the need for translation, but must be designed carefully to ensure they are easily understood.

Source: From *Assembly Instructions, Weber Spirit Grill E/SP 210/310,* Weber-Stephen Products Co., Palatine, IL, 2007. © 2007 Weber-Stephen Products Co. Used with permission.

LAYING OUT THE PAGES

At the beginning of this section, we recommended that you plan the graphics at the same time as you plan the text, using storyboards or some other similar system to ensure that the two modes of communication develop together. This alone will go a long way to ensuring that the graphics in the manual work well with the text. Graphics, like text, often convey information most effectively in a general-to-specific order. Just as you would begin a new section of text with an overview, you can provide an overall illustration of an assembly before you present detailed drawings of each component.

When you begin to lay out the pages, you will appreciate having planned text and graphics together. Regardless of the format of your manual, a few basic principles apply:

- Make sure graphics and text follow the same format.
- Always refer to the graphic in the text.
- Place the graphic as soon as possible after the first reference to it.
- Avoid making the reader flip back and forth between two pages.

Chapter 4 cautions against allowing a graphic to "leak" into the gutter between columns of text. But what if you have a two- or three-column layout and you want to ensure that your graphics are big enough to see clearly? It's perfectly acceptable to size a photograph or drawing to span more than one column of text—you just have to be sure that it covers the full width of the columns. In other words, a graphic that is two-columns wide is fine; one that is one-and-a-half-columns wide is not. The former maintains the integrity of the two-column format. The latter simply makes the page look disorganized and unattractive.

Whenever you use a graphic, you must let the reader know why it is there. Always refer to the graphic in the text—it is quite disconcerting to come upon a figure without explanation (particularly if it has no caption other than a figure number). If you have planned for the graphics from the start, this will be easy and natural to do. As you draft the text, you already know in general terms what visuals you plan to use, and it will almost be automatic to refer to them. If you have not planned text and graphics together, it will be more difficult. You will need to make a conscious effort to go back after the graphics are chosen and insert appropriate references to them.

If seeing "Figure 5.24" with no explanation is disconcerting, reading a reference to "Figure 5.24" and then being unable to locate it is downright irritating. The ideal location for a graphic is right after the first reference to it—but not before. If you make the reader thumb through page after page to find the figure, it won't take long before the reader decides it's not worth the effort. If that happens, including the visual has become pointless. Even if it "costs" an extra page to accomplish, it may be worth it to place a graphic near the reference to it.

Avoid making the reader flip pages back and forth unnecessarily. For example, avoid putting a parts diagram on one page and the callouts on the overleaf. It is far better, if the drawing and the callouts will not readably fit on one page, to place them on facing pages. Remember that your readers are often looking back and forth between manual and product—making them also turn pages is too much. At all costs, make sure that drawing and callouts are oriented in the same direction. In other words, do not make your readers look at a "landscape" drawing and then turn the manual to read a "portrait" parts list.

Naturally, the "perfect" layout is seldom possible. As with so much else in manual production, laying out graphics and text requires balancing the ideal of perfection against the reality of time and budget constraints.

SUMMARY

We live in an increasingly visual society. The growth of the Internet and widespread use of video have shifted the balance from text-based sources of information to visual-based ones. Additionally, the growth of global commerce has made visual communication in manuals ever more important as a way to reduce the cost of translation. Even without these shifts, the visual elements of manuals help users understand products. Just as white space and headings help readers visually explore the organization of the manual, graphics help readers explore the product. To be effective, however, graphics should be planned from the beginning and developed along with the draft of the manual text. The goal is for graphics and text to work together to convey information more effectively than either could alone.

You should consider using a graphic of some kind whenever you have a picture in your mind as you plan and write the text. Some situations cry out for graphics: when there is a lot of quantitative information that would be difficult to understand in paragraph form, and when the user must perform a procedure. It's usually easier to show someone how to do something than to tell how to do it. As noted, translating manual text can be expensive, so if your products are sold overseas, leaning toward visual communication rather than verbal is usually a good idea.

These days, many different types of graphics can be easily and relatively inexpensively produced. The development of computer-aided drawing and digital photography has made a world of difference in the ease with which good illustrations can be made. However, the writer still has the responsibility to choose the appropriate type of illustration, whether it be a photo or a drawing, a table, or a chart. Which you choose depends on what you're trying to show; all have strengths and weaknesses for particular purposes. Whatever kind of graphic you decide on, take care to ensure that it is designed and produced for easy use. Make sure that graphics are large enough and clear enough to take in the basic idea at a glance. Unlike text, visuals are seen "all at once." Make sure your users will easily grasp the point of the illustration.

Finally, make sure that your graphics and text work together. If you design them together, they probably will; if you slip a few illustrations in as an afterthought, they may not. Tie the two together by reference within the text and by designing your page layouts for easy use.

The digital age has changed the way we view manuals. As the next chapter shows, it's not just that the books contain more pictures—sometimes they aren't even books anymore.

CHECKLIST: EFFECTIVE GRAPHIC DESIGN

- Have I identified all the places a graphic might help my user? Have I looked at the product as a first-time user?
- Have I planned my graphics and text together?
- Have I chosen the best type of graphic for what I want to show?
- Are my photographs shot from the best angle? Retouched as needed?
- Are my drawings simple and uncluttered? Appropriate to the audience?
- Have I eliminated unnecessary gridlines and other clutter from my tables?
- In tables, are the data arranged vertically for easy comparison?
- Are all the graphics—and elements within them—big enough to see easily?
- If drawings have been reduced, is any text still readable?
- Have I called out parts clearly? Is it obvious what each callout refers to?
- Is every graphic referenced in the text and placed close to the first reference?
- Are the graphics integrated well into the page design and layout?
- Are the graphics as clear, inviting, and easy to use as the text?

NOTES

1. Ralph Lengler, "Identifying the Competencies of 'Visual Literacy'—A Prerequisite for Knowledge Visualization," *Proceedings of the 2006 Information Visualization Conference of IEEE Computer Society*, 2006.
2. James R. Flynn and William T. Dickens, "Heritability Estimates versus Large Environmental Effects: The I.Q. Paradox Resolved," *Psychological Review* 108, no. 2 (2001): 346–369, cited by Lengler, ibid.

6 Thinking Outside the Books

OVERVIEW

When we think of an operator's manual, we usually envision a paper document filled with words and some pictures—maybe a three-ring binder, maybe a few pages stapled together, maybe even a single sheet of instructions. Increasingly, "manuals" are taking on other forms entirely. One effect of the digital revolution has been to make it possible for people to access information on demand. With access to the Internet and a few clicks of a mouse, a user can download a manual, watch a video of a product in use, view an animation that explains how it works, and perhaps even chat with a technical support specialist in real time. Some products even have the technical documentation built right into the product—there is no manual apart from the product itself. At the same time, as audiences have become more diverse, the need for specialized manuals has also grown. With all these choices, what's a writer to do?

This chapter looks at some of those new ways to communicate information to users. Even though the medium changes, the same principles apply. Whatever form your communication takes, you must make it appropriate for the audience and as easy to use as possible.

PUTTING IT ON THE INTERNET

The advent of computers dramatically changed the way we create manuals, but the changes extend beyond the improvements in the technology used to create paper documents.

THE DIGITAL REVOLUTION: BEYOND PAPER

A few years ago, it was widely predicted that the digital revolution would create a paperless society. That prediction was most certainly wrong. Global consumption of paper continues to rise, but even if we are not using less paper, we are certainly using other media in addition to paper. Digital information, especially, has seen spectacular growth. In a few short years, the Internet has become omnipresent and indispensable in every corner of the world. While access is often more expensive and less convenient in developing countries than in the developed world, it is available clear around the globe from Afghanistan to Zambia. Access will only increase. Not only will global commerce and scientific collaboration push broader Internet access, but enterprises such as the innovative One Laptop Per Child project[1] will ensure that coming generations experience access to the Internet as more available than even access to electricity and clean water.

Just as the computer made "desktop publishing" available to everyone (with mixed results, to be sure), the advent of smartphones and inexpensive digital video cameras has made video technology available to all. Combine these with Internet access and you have a powerful new platform to train users—or confuse them.

INFORMATION ON DEMAND

As noted in Chapter 1, we have become an "on-demand" society. It stands to reason that product users are likely to have the same expectation of information at their fingertips whenever they need it. Fortunately, new technology makes it ever easier for manufacturers to meet that expectation. The digital revolution offers a multitude of new avenues to help customers learn how to use products safely and to trouble-shoot problems—all available 24 hours a day, 7 days a week, and in a variety of languages.

The simplest of these options is to make a copy of product manuals available on the company's website. Many manufacturers already do this, usually in a .pdf (portable document format) file. If the manual is produced digitally, saving it to a .pdf format is a simple procedure. Older manuals must first be scanned into digital format, but that, too, is a simple procedure.

TECHNICAL ASSISTANCE: A MOUSE-CLICK AWAY

The Internet offers another resource for manufacturers and customers: training in how to use and maintain the product and technical assistance in troubleshooting. The Internet is a very flexible medium. It can store documents, video, music, voice, and animations. Companies are beginning to make use of this versatility to provide information to customers above and beyond what is typically possible in a paper manual. For example, a troubleshooting section in a manual takes one of two forms: either a table, listing problem—cause—solution; or a procedure, sometimes presented as a flowchart. In either case, the manual tells the reader what to do. With the capabilities of the Internet, the company can instead show the reader what to do.

Nor does showing the user require professionally produced video. Sometimes an animated sequence is even more effective because it can isolate the desired sequence and show things that a camera cannot, such as the movement of fluids. Animations also allow showing of movement (a cap being unscrewed, for example) without the need for a human hand to cause the motion—and get in the way of a good picture.

The possibilities are limitless. You could combine photos, animation, and sound to demonstrate normal operation and abnormal conditions, such as a fan out of balance or a squealing bearing about to burn out. You could show the flow of product through a processing machine. You could demonstrate technique for cleaning or sharpening a blade or tool. In fact, an Internet site can be the next best thing to having another person show the user how to do a procedure.

For all these Internet-based options, from copies of manuals to troubleshooting and training, some companies charge a fee for access and others offer the material free. Whether to charge or not is, of course, an individual decision. Two things to consider in making that decision:

- Are you being penny-wise and pound-foolish?
- Are your users likely to pay for access or will they simply not use web-based material (or use unauthorized alternatives)?

Both these questions can be answered with a little research. Let's look more closely at them.

Penny-Wise and Pound-Foolish

The phrase "penny-wise and pound-foolish" simply means avoiding a small cost and not recognizing that doing so may carry a much larger cost. It's like driving 20 miles out of your way to save two cents per gallon on gasoline—you burn more money on the trip than you save with the cheaper gas. If you charge a fee for accessing information—particularly safety-related information—you can be assured that some people will choose not to pay the fee, and therefore will not have access to the information. The fees received may help pay the cost of producing the web-based information, but if lack of access results in someone's being injured because of being unaware of some safety issue, the manufacturer may end up paying out a lot more than it received in fees. A better approach, if the company wants to charge at all, is to offer operator manuals and safety information free, but charge for supplemental material.

Will Users Pay?

You can find out whether your expected users are likely to pay a fee, and if so, how much of a fee, simply by use of surveys, focus groups, and other standard marketing tools. If they are not willing to pay a fee—or not willing to pay very much of a fee—the company might be better off giving access to the material without cost. After all, if the customers won't use the site because they have to pay, then all the development cost is wasted money. On the other hand, giving the information away may forestall calls or emails with questions to the company's technical assistance number. The answers to these questions will vary with the product and the users. In general, products that are highly technical and require specialized knowledge to use or service can command a higher fee than simpler or more familiar products, particularly if alternative sources of information are available.

Even if users would be willing to pay for the information, the company may decide to give it away to foster good public relations. Making product information widely available, especially safety-related information, makes a strong statement that the company values its customers' well-being—and may result in increased future sales as well as fewer service calls.

INTEGRATED PRODUCT DOCUMENTATION: ARE PAPER MANUALS TODAY'S BUGGY WHIPS?

Remember software manuals? In the early days of broad-based applications like word processors and spreadsheets, the actual software (on floppy disks) was inevitably accompanied by one or two manuals. Often there was a user's manual and a much thicker reference manual. Not anymore. Now you may get a CD (although most software providers have gone to a licensing business model, where you pay a recurring fee to download and use the software, but cannot buy it), but no book. The manual may be on the disk, or more commonly is built right into the software as "context-sensitive" help. A similar transformation is beginning to happen with other products as well. For example, your office copier may have a built-in menu of instructions for how to choose features, such as two-sided copying. It may display a graphic that shows the location of a paper jam, and may even show an animation of how to get at and remove the jam. The robotic vacuum cleaner Roomba[2] offers a built-in demonstration, with a recorded voice explaining how it works. The same recorded voice provides troubleshooting solutions, asking you to "Please clean Roomba's brushes" when a thread has become wrapped around a brush so it cannot turn.

Are paper manuals obsolete? Even if a product is not so far advanced as to have user information built into it, would it be OK just to post the manual on the Internet and dispense with the book altogether? It would certainly be a lot cheaper, and could guarantee that only the most up-to-date information was available. You wouldn't have to worry about how to jam a dozen language translations into one book without making it totally unusable—just post multiple translations on the web. Internet-based manuals never wear out, never get dog-eared, never get spattered with oil or smudged with dirt. They don't get lost and they aren't forgotten when the product is sold. With all those advantages, why do we still have paper manuals at all?

One reason is that paper manuals are sometimes more convenient and user-friendly than a web-based manual. For example, some products are routinely used in places where a computer with internet access is unlikely to be convenient. Consider these: an outboard motor on a 14' fishing boat, or an air compressor on a construction site, or a commercial dishwasher in a restaurant kitchen. Even if a computer is readily available, the sheer size of the manual may make using it online inconvenient. Reading long documents on the screen is fatiguing. It's not as easy to write notes on the pages or bookmark frequently used information as on a conventional paper manual. The computer may be located in a different area from the product. Remember that people often will look back and forth from manual to product when they are performing a procedure—that's hard to do when the two are in different rooms.

Not everyone has access to the Internet. Even if access is available, it may be slow or intermittent, even in the United States. Downloading a large file may take minutes or even hours. The less convenient using the manual is, the less likely people will be to bother. While it is certainly an excellent practice to post information on the web, including product manuals, for many products it is not yet practical to dispense with the paper version altogether.

SHOULD WE PUT ALL OUR MANUALS ON THE INTERNET?

Even if your product is accompanied by a paper manual (or more than one), also making product manuals and other documentation readily available on your company's website is an excellent idea. The goal is always to make it as easy as possible for your customers to get the information they need. For consumer products, you have no guarantee that users will keep the manual—or be able to find it even if they do keep it. For many of us, the filing system for product manuals is the "junk" drawer in the kitchen. For commercial and industrial products, the manual may be kept in an office that is not always accessible to users, or the product may be used in the field at an outlying location. It's much more likely that someone will have a laptop with Internet access on a jobsite than that every relevant product manual will be brought along to the site.

Another reason to put manuals on the Internet is that many products are resold, often without the manual. Consumer products like lawn mowers and pressure cookers are bought at garage sales, commercial and industrial products like wood splitters and punch presses are often purchased at auction or acquired as part of real-estate transactions. In any case, it's unlikely that the original manual will go along with the product. Having the manual available online ensures that users will be able to read important instructions and warnings.

Sometimes manufacturers worry that putting manuals online for products no longer sold will expose them to liability, since the product may no longer meet current design standards or the warnings in the manual may not be up to date. They think that if they make the manual available, they are encouraging use of a product that is, by today's standards, not reasonably safe. They worry as well that if they make the manual available and there is a lawsuit, the plaintiff's attorney will be able to use the manual to attack them in court.

Instead of asking whether old manuals pose a liability exposure, ask instead whether old manuals are better than no manual at all. Remember, the manufacturer has a duty to provide reasonable instructions and warnings. Putting old manuals online is better than the alternative for three reasons:

- The fact that newer products have more safety features does not automatically make an older product unreasonably dangerous.
- Without a manual available, many users will try to figure out on their own how to operate the product—with potentially disastrous results.
- Putting the manual online offers an opportunity to update warnings and highlight procedures to mitigate known safety issues.

Even if there is an injury and a lawsuit, the company will be able to show that it made a good-faith effort to get vital information into the hands of users, even when it had no way to identify or locate those users.

INSTRUCTIONAL VIDEOS AND WEBSITE ANIMATIONS

With the advent of home VCRs, some manufacturers started to use videotapes to supplement or substitute for installation and instruction manuals. Tapes gave way to DVDs, and now videos are streamed over smartphones and tablets. The medium has changed, but the idea is the same: if what people really want is to learn from another person rather than to read a manual, why not give them exactly that? Instead of making them read a description of what to do, why not show them how to do it by having them view a video of someone performing the procedure? It's almost as good as having an expert in the room, right? The answer, of course, is that it depends. Instructional videos can be very effective—or they can be disasters. What makes the difference?

What Makes a Video Work?

How effective a video can be depends on three factors:

- Type of product
- Technical quality of the video
- Planning and scripting

When all three work together, the result can be splendid. When one or more goes awry, it would be better to stick to paper and ink.

Type of Product

Not all products lend themselves well to using a video to teach about them. As an extreme example, a video about how to use a word-processing program or hook up a surround-sound system would probably not be very effective (although an animation might work). On the other hand, a video showing how to extend and level the outriggers and operate a boom lift (commonly called a "cherry-picker") could be very effective. What makes one type of product work in video format and another not? Generally, products well-suited for video instructions have two characteristics:

- Moving mechanical parts
- Scale that allows parts to be easily seen in context

The word-processing program and the surround-sound system do not have a lot of moving parts to show. Typing fingers and letters appearing on a virtual page are not going to keep anyone's visual interest, nor are a succession of shots of a hand plugging in cables. On the other hand, watching the hydraulically operated outriggers extend and level the base of the boom lift and watching the bucket itself rise and maneuver would be visually engaging. If you are considering using video to help teach about your product, ask yourself where it fits on the spectrum between word-processing and boom lift. Would there be enough visually interesting parts to carry the video? As a general rule, the more mechanical the product, the better it is for a visual medium. The more electronic or software-based, the worse it is. Many products today are a mix of both. Your task as product documentation specialist is to evaluate the mix and decide if there would be enough visual interest to make a video worthwhile.

The other important question is how easy it will be for the viewer to understand the location of particular parts in the context of the whole product. If the actions to be shown involve very small segments of the product, a video might not be useful. For example, if you have to show an extreme close-up to make the "action" visible for a procedure (attaching wires, turning an adjustment screw by small increments), the viewer may lose the perspective of where this takes place in the context of the entire product. If the parts are very small, it is often difficult to show them without the technician's hands in the way as well. The size of the product itself is not the critical variable—a very small product can work well on video, because the video can show the whole thing at once. Similarly, a very large product can work well. The difficult mix is a large product with very small-scale procedures.

Technical Quality

We've all seen jerky, out-of-focus, poorly edited home videos. Just as the common availability of desktop-publishing software puts page design in the hands of those who have no training in how to do it well, smartphones make movie directors of us all. Shooting and editing good-quality videos are not tasks for amateurs. Professional videographers know how to choose and manipulate lighting, depth of field, camera angles, wide and tight shots, and other aspects of the filmmaker's craft to get the desired effects. Above all, they know how to plan for editing the video as well.

Editing the video is every bit as important as the initial video shoot—and it can be even more time-consuming. In a well-edited video, the watcher is generally unaware of the transitions from scene to scene and from one camera angle to another. In this way, video editing is much like editing prose—if you do it well, the user never notices the cuts and pastes. To achieve good results, an editor must have sufficient quality and quantity of video to work with. The best editor cannot make a polished final product if the raw footage is underexposed or out of focus or he or she doesn't have enough "footage" to work with. It's not uncommon for a professional to shoot six hours of video to produce a 20-minute finished product. That six hours of video is made up of shooting the same things from multiple angles and shooting the same sequence of actions in different ways. When it comes time to edit the video into a final product, the editor has lots of choices of how to present the information. Unlike a document editor, who can simply move words around and insert new phrases of his or her own devising, the video editor is generally limited to what's already been shot. In general, it doesn't work to go back and re-shoot some sequence, because try as you might to recreate the exact conditions of the original shoot, something will be different. The technician's hand will have a scratch that wasn't there the first time, or the light will be different, or the shop floor will have been cleaned and waxed. These minor differences make it very difficult to splice scenes together without continuity problems.

You (or your boss) may be tempted to send a couple of people out with the company videocam in order to save money. In the long run, spending the money up front for a professionally produced video would almost always be a better choice. A poor-quality video, like a poor-quality manual, will turn customers off and reflect badly on the company. A well-produced video, on the other hand, will serve your customers well and convey a professional image of the company.

Planning and Scripting

Developing a good instructional video, like writing a good manual, requires a great deal of planning. Each scene must be planned in much the same way that you would plan an outline for a book. What is the scene going to show? What's the best way to show it? Should the camera show the technician or should the scene be shot from the technician's point of view? What words will accompany the video? Do you want to record the sound separately as a voiceover or have the technician speaking? Or some of each? Having the technician speak may be more "friendly" than a voiceover, but doing so makes correcting a flubbed line much harder. Because sound and picture are linked, you cannot easily re-record the voice and dub it in.

PLANNING AN INSTRUCTIONAL VIDEO

When you plan an instructional video, three aspects can be especially difficult:

- Controlling the pace
- Allowing for updates
- Making it useful as an ongoing reference

Using a linear visual medium makes all of these more problematical.

Controlling the Pace

While as yet relatively little research addresses how best to present procedural instructions in a video, some has been done in the field of instructional design and distance education. A key finding of this research is that "interactive" videos, that is, videos that allow the user to control the flow of information, are more effective than those that require the user to watch the video passively from beginning to end. How do you make a video interactive? By employing the same principles as in manual design. We know users won't read a manual from beginning to end, so we find ways to make it easy to navigate to what they need in the moment. In a manual, this means using headings and references to get the reader to the right spot quickly. In video, this means making it easy to start and stop the video, to search and browse for information without having to fast-forward through the entire video, and to play and replay a selected portion.

   Without question, the most common flaw of product-related instructional videos is setting too fast a pace. With a paper manual, you have to plan for the user to look back and forth between manual and product. You might think that with a video, since you are showing the product on the video, the user has no need to look back and forth. In reality, the user has exactly the same need to check where things are on the product. A video may be a "movie" while a photo in a book is a "still" shot—but neither one is the actual product. Users will want to make sure that they understand what to do by locating parts on their own product.

   How do you do this in a video? A book will wait patiently for the user to come back and move to the next instruction, but the nature of a video involves ongoing action. One option, of course, is for the user to repeatedly hit the PAUSE button. This option may not be desirable if it results in the loss of the smooth flow of the narration. Another option is to plan for these breaks. Either show a small segment of action and then direct the user to pause the video to locate and perhaps perform the action on the product, or simply plan for supplemental narration to take up the time when the user is looking at the product. Either way, break up the action into small segments, much like breaking up a long procedure into sections. Without such planning, the procedures shown will speed past way too quickly for the viewer to apprehend. Like a magician's audience watching a sleight-of-hand trick, a user watching a too-fast-paced instructional video simply cannot keep up with the action.

Allowing for Updates

Chapter 3 suggests designing a manual in modular sections to make it easy to update—simply pull out the outdated portion and plug in new text and illustrations. To a degree, you can do the same thing with videos, but it's a bit more difficult. You will need to build segmentation in from the initial planning, to avoid choppy and distracting transitions. Noticeable changes will be impossible to avoid entirely, but you can minimize them by following these guidelines:

•   Identify what is likely to need updating and plan for it.
•   Minimize showing a technician's face and body—focus on the hands.
•   Keep the visual background neutral and clear of extraneous objects.

If you plan ahead for updating, you can put deliberate breaks into the video before and after segments that might need to be replaced. For example, you can switch from a view of the product to a view of the narrator. That way, when you replace a segment, you do not have to try to smooth over a choppy splice. Or you can use some other neutral interlude to divide segments. While you must plan for segmentation, beware of using so many breaks that you destroy the flow of the video.

In any scene showing a technician performing a procedure, whenever possible, focus on the hands rather than showing the full figure or face of the technician. Doing so gives two advantages. First, a close-up of the hands is more likely to show the details of the procedure better, and second, if you have to update that segment using a different technician, the change won't be as distracting. While hands certainly vary from person to person, the variations are usually not as noticeable as if the technician suddenly morphs from a muscular 25-year-old to a balding middle-aged man with a paunch. Similarly, fashions in clothing and hairstyles tend to change rapidly, sometimes even more rapidly than product models do. Hands today look pretty much like hands did twenty or thirty years ago, with the exception that today they may bear colorful tattoos. Like clothing and hairstyles, fashions in body art also change. A few years from now, tattoos may be out of favor.

Keeping the visual background as neutral as possible will make substitutions less noticeable as well. If earlier shots of the product show a pile of lumber or a trash can in the background, but suddenly in the next scene it disappears and a workbench magically takes its place, the viewer will be distracted, even if not entirely sure why. Keeping the background neutral and uncluttered also makes it easier to focus on the product and the procedure you want to show. The viewers will be more able to pay attention to the instruction, without having their attention divided between the product and background clutter.

Video as an Ongoing Reference

Video is a linear medium. Like music or a written narrative, it relies on sequence to make sense. You cannot as easily skim a video as you can a chapter of a manual. Nevertheless, just as users reading a manual do not want to have to read the entire book just to look up a specific piece of information, viewers do not want to have to fast-forward through an entire video to find the segment they need. Even if the video is divided into scenes that one can access in any order, it's still a linear medium, and linear media work best as sequential information, not random-access. You wouldn't want to put a dictionary on a video. For this reason, companies often use videos for very specific one-time or occasional activities, such as installation or periodic maintenance procedures, and provide a conventional manual for later reference.

Safety Information in Videos

Videos can also be effective to convey how to safely operate the product. An example is a video showing the proper way to operate a tractor on a slope or to secure a load to a hoist. However, as yet there are no standards and little research on how best to convey safety warnings in a video. If you look at various product videos, you will find different approaches for conveying warning and hazard information:

- Voiceover narration
- The technician in the video stating the hazards and warnings
- On-screen text (crawls at the bottom of the screen, warning-label overlays, bullet points)

All of these can be effective, but the visual image presented is crucial. Just as people respond more to body language than to the actual words spoken, the visual images presented in a video carry more weight than any spoken or printed warnings.

Consider the ubiquitous TV commercials for pharmaceuticals. Federal law requires that drug manufacturers include relevant warnings regarding side effects, potential drug interactions, contraindications, etc. in advertisements, but the manufacturers of course also want to portray their products in a positive light. Nearly always in these commercials, the required warnings are presented near the end using a voiceover—while the images on the screen are of smiling, healthy people enjoying an active life—thanks to the medication. The voiceover may be recounting perfectly dreadful possible side effects, but the visual image is all positive. The manufacturers hope that the visual images of good health will outweigh the audio messages of scary side effects.

Unlike pharmaceuticals, most products do not have side effects, so manufacturers are not faced with the problem of how to entice people to buy and use the product in spite of the potential for negative outcomes. Instead, the challenge is more straightforward: how to ensure that people use the product safely. In this respect, the goal is the same as with a paper manual—to keep the user safe. Whichever technique you use, be sure that the images on the screen do not conflict with the message. For example, if the on-screen text warns that you must always wear eye protection while using the product, do not show someone using the product who is not wearing safety glasses. If you have the on-screen narrator verbally stating a hazard and how to avoid it, have the person simultaneously *doing* the avoidance action. The visual image will reinforce the verbal message.

You Can Find Anything on YouTube

According to a 2018 report by the Pew Research Center, 73% of Americans use the video-sharing site YouTube—more than any other social media platform, including Facebook.[3] Of course, many of the videos shared on YouTube feature cats or musical performances (or even cats performing music), but a good number of them show either product uses and features or repair procedures. Many manufacturers have their own YouTube channel to promote their products, provide video answers to Frequently Asked Questions, and take users on virtual tours of their company's products and processes. YouTube is also an excellent medium for how-to videos showing users how to perform various procedures from set up to repair.

You can find out how to do almost anything on YouTube, from repairing a hole in carpet to replacing the high-temperature limit switch on a furnace, to maintaining a pellet mill. The challenge is that not all of the videos providing instruction are posted by manufacturers or even qualified repair technicians; some are posted by do-it-yourselfers. Some of these "unofficial" how-to videos are very well done and show proper procedures done safely. Others not so much. For example, if you search for videos on mounting a tire on a rim, you will find numerous examples. Some show the procedure done safely using a tire machine and safety cage; others show the procedure being done at home, using prybars and brake fluid—or worse, using an ether explosion to mount the tire. Your own product videos may show the operator wearing appropriate PPE (safety glasses, ear protection, and gloves) while operating your product, but your product may also be featured in unauthorized videos in which the subject does not wear any safety equipment or follow safe procedures.

What can you do? While you may not be able to prevent someone from posting a video featuring your product, you should make every effort to be aware of what's out there. If you find unauthorized videos that promote unsafe practices, you may want to use your own YouTube channel and your company website to promote to safe procedures and warn against unsafe ones. If you find a number of instances of specific safety procedures being disregarded, try to determine the reason. Often the reason that people ignore safety procedures is that they make a task inconvenient or slower. Look at ways to change the design or the procedure to make doing the job safely the easier choice.

WEBSITE ANIMATIONS

Many companies use animations on their websites to promote their products (see for example, www.emhartglass.com or www.deere.com). Some of these animations simply identify key features that the company wants to emphasize or show how the product can simplify or automate a process to increase efficiency and save money. Some of them, however, also show how the product functions. These "explainer" animations, unlike photographic videos, can easily show processes that are normally hidden from view.

Website animations can also be useful to provide instructions for installation or repair. Instead of a video showing a technician's hand removing the screws holding an access panel in place, an animation can show the screws unscrewing themselves and the panel moving out of the way unaided. What's the advantage of animation over video? For one thing, you don't have to worry about the technician's hand being in the way of seeing what the procedure is. Additionally, with an animation, as with a still illustration, the artist can remove unneeded clutter, use color to highlight parts of interest, and not have to worry about proper lighting and exposure. The drawback is that animation can be quite expensive, because it is usually labor-intensive. On the other hand, professional video production is also expensive. As technology develops, animation is likely to become more affordable, particularly with the advent of animation software that uses computer-aided design (CAD) drawings of machine parts, much as architectural building information management (BIM) software uses computer-drawn blueprints to create three-dimensional architectural views. These architectural renderings show not just the finished design, but by selecting various layers, you can virtually peel back the walls and see the plumbing, wiring, and HVAC layouts as well.

One current example of CAD-based animation is BILT®, which is an animation service for manufacturers. The manufacturer sends a sample product and CAD information, pays a fee, and BILT produces animated assembly instructions. These are available to the consumer or technician free of charge, using the downloadable BILT app. The company can direct the consumer to the BILT app in the manual or on the product packaging.

Like video, animations need to be well-planned and executed. And like video, they require that the user have access to a computer, tablet, or smartphone at the time the information is needed—and that the lighting conditions are such that the screen is viewable. A farmer in a cornfield on a sunny day may find a paper manual more useful than a web animation, even if a device for viewing it is handy.

PICTURE THIS: THE NO-WORDS MANUAL

Translation is expensive. Words can be interpreted in different ways by different people. Why not simply dispense with them altogether and make the paper manual completely visual? A picture-only manual can be effective for certain kinds of products and procedures, but it does not work for everything. Surprisingly, producing a pictorial manual requires many of the same steps as producing a conventional manual. Nor are pictorials necessarily more precise than words.

PICTURES ARE ABSTRACTIONS TOO

A verbal description of a product part is quite removed from the actual part. It is an abstraction and therefore requires training to interpret it. Before being able to interpret a written description of, say, a splined shaft, a person must (1) understand English, (2) be able to read, and (3) know the meaning of the term "splined." Wouldn't it be simpler just to provide a drawing?

Certainly, one need not know English (or any other specific language) to interpret a drawing, but one still does need some training. Visual communication has its own rules and conventions, just like language, and these are not always intuitive. For example, in manual illustrations, close-ups are often shown as an inset into a larger drawing, with an arrow and a circle or rectangle showing the location of the enlarged portion. We show hidden portions of a product with dotted lines. We show how parts are assembled by lining them up in an "exploded" view and running broken lines through the centers of each part. These are all conventions that we have become familiar with, but they are still learned interpretations. Some illustrations are even more abstract. Consider, for example, the engineering drawing of the valve shown in Chapter 5 (Figure 5.5). Learning how to read that requires considerable training.

Even a perspective drawing involves conventions. We show two items of the same size as different sizes when one is farther away. We show lines and planes that are in fact parallel (railroad tracks, walls) as converging as they recede in the distance. We use a few diagonal parallel lines to indicate a reflective surface such as glass. We draw just a few bricks to suggest that an entire surface is bricked. We use arrows to show the direction of motion (a curved arrow above a knob to show it should be turned clockwise, for example). Most of us don't think about these—we learned them as children and more or less take them for granted. They are no less real for that. Look back at Figure 5.23. Notice how many separate visual conventions are used in just that small section.

KNOW YOUR USER

The fact that pictorial-only manuals involve learned conventions just like word-based ones simply means that the designer must be aware of that fact and plan accordingly. Just as a writer needs to learn about his or her audience, the designer of a graphics-based manual also needs to know the user.

User analysis for a pictorial manual is essentially similar to user analysis for a word-based manual. The difference is that once you have identified your users and gauged their familiarity and comfort with the product, you need to shift toward learning about their visual literacy. Here you may find significant cultural differences. In Japan, pictorial-based manuals are fairly common, and the typical user is quite comfortable with that style of instructions. In North America, they are not common, and some users may be frustrated by the lack of words. Europeans are more used to symbol- or pictorial-only communication in general (warning labels, signs, and so on) than are those of us on this side of the Atlantic, probably as a result of the many different languages in common use.

As the world becomes more and more of a "global village," and visual media become more and more widely used in all arenas, it is likely that those in the United States and Canada will become more at ease with pictorial manuals. Until then, you may have to build into your manual some training in using a visual-only format.

PLANNING THE PICTORIAL MANUAL

Planning a pictorial manual is similar in many ways to planning a conventional word-based manual. However, not all kinds of information work well in pictorial mode, so the first step is to consider whether you should use a pictorial manual at all. Generally, these types of information can be effectively conveyed graphically:

- Assembly and disassembly instructions
- Adjustment or repair procedures that involve physical actions
- Certain safety-related information

Assembly and disassembly instructions, such as the segment shown earlier in Figure 5.23, can often be shown in pictures alone, sometimes supplemented by words. In fact, showing assembly and disassembly in pictures only is a good deal easier than trying to describe them using only words. For such instructions to be useful, however, you must be sure that the drawings or photos are large enough and clear enough to be easily understood. If a full-size computer-aided design (CAD) diagram is reduced to fit a standard 8½ × 11-inch page, critical detail may be lost. Without words to supplement the visual information, it may not do the job.

Adjustment or repair procedures can often be shown pictorially. Correct alignment or adjustment of mechanical parts can be illustrated by a series of drawings or photos showing both proper and improper placement. Photos can indicate—often better than words—what a part looks like when it is "excessively" worn, corroded, or otherwise compromised and should be replaced. Illustrations can show correct placement of tools, optimum grinding angles, or positioning of a technician's hand. Not all procedures work visually, however. If the key indicator that the procedure has been done correctly is an audible sound, or a tactile sensation such as a shaft turning "smoothly," that is difficult to convey with a picture alone.

Some kinds of safety information can be shown in pictures. In fact, it is common in the United States and Canada to use pictorials on warning labels to show the hazard, and in Europe to convey the entire warning label content. Pictorials can show how to operate a product safely by showing correct practices (and unsafe ones), by indicating danger zones, and by illustrating what happens when safe operating procedures are not followed. Sometimes safety information requires more detail and explanation than can be conveyed strictly visually, but often at least the specific hazard can be shown in a picture, if not all the nuances.

HOW DO YOU KNOW IF YOU'VE SUCCEEDED?

Part Two of this book has focused on how to create manuals that meet users' needs, including the real but unstated need *not to feel stupid*. How can you tell if you've succeeded? Do you just take your best shot at applying the principles and techniques presented and hope the manual works? Companies have long employed usability testing to ensure that their products function as intended. Usability testing can also be used to see if the manual functions well, whether it's text-based or pictorial.

USABILITY TESTING METHODS

Just as with hazard analysis, methods for conducting usability testing vary, from simple to complex.

Here's a sampling:

- Informal and In-House
- Beta-Site Testing
- Formal, Systematic Analysis

Informal and In-House

Some companies invite employees from other divisions, friends, family members—anyone who is a true stranger to the product—to "walk through" the manual, following its instructions and descriptions. Technical publications staff stand by and listen and watch, but they don't provide verbal backup to the manual unless the user asks for help. Wherever they have to break in, explain more fully, or provide more information, the manual probably needs revision or clarification. This option is very simple and inexpensive, but can provide valuable feedback.

Beta-Site Testing

Some companies, including those doing military contract work, have selected locations called *beta sites* used for testing products and debugging manuals. At beta sites, users are asked to perform the operations described in the manual. For instance, an airplane mechanic may be asked to follow the manual for installation of a new landing gear. All difficulties and snags experienced by the user in following instructions are monitored and recorded. The manual is corrected, revised, and retested to assure that instructions are clear. This approach is more complex and more expensive, but yields more specific information, and may be required in a military contract.

Formal Systematic Analysis

Some companies dedicate considerable money and time to formal usability analysis. The setups can be as simple as the use of a video camera with sound capability to record the manual user at work. More elaborate user feedback setups place users at a desk equipped with a microphone, the manual, and multiple video monitors. Manual pages are presented, and as users work their way through tasks, their activities and voices are recorded. Design engineers, safety engineers, and technical publications staff monitor the users at work from behind one-way glass. The video record of the user at work is digitally synchronized with video of the manual pages. The combination of voice plus video of both user and manual page allows observers to know exactly what the users saw on the page, what task they were trying to perform, and what they said and did about it.

Other Alternatives

Many companies ask users for feedback on their product and manual, either by including a postage-paid response card in their manuals or by sending an email survey. Most people are feeling "surveyed to death" these days, however, so the response rate to either method is pretty low.

USABILITY TESTING FOR VISUAL MANUALS

Even more than with word-based manuals, usability testing is critical with visual manuals. While people can certainly differ over the interpretation of words and phrases, it is likely that most literate people will agree on the meaning of text in a manual, simply because we communicate information using language all the time. The result is that most people are fairly well-trained in correctly interpreting straightforward prose. Most people are much less skilled in interpreting visual information, so it stands to reason that the potential for misinterpretation is much greater.

Like other usability testing, your test subjects should be similar to your expected users. If your product is a surge arrestor used in the electric power industry, you would expect the users to be able to correctly interpret standard circuit diagrams. On the other hand, if your product is an automotive fuse, you could not make the same assumption, as your users could range from trained mechanics to automobile owners with no electrical knowledge whatsoever. If your product is sold internationally, make sure you test the visuals against international subjects as well.

In addition to determining whether a pictorial manual is effective overall, usability testing can indicate whether there are any parts or aspects of certain visuals that are confusing or ambiguous. As with word-based manuals, you may never get 100% agreement, but if a substantial segment of your test subjects find a specific drawing or photo confusing, then it should be redone. One of the simplest ways to test visuals is to ask the viewers to describe in words what the visual is showing. You can then match the verbal descriptions against the intent of the visual to see whether it works as planned.

SUMMARY

Technology has made the traditional paper manual only one avenue through which a product user can gain information. As we have increasingly become an on-demand society with the expectation of access to needed information whenever and wherever we want it, manufacturers have responded in many ways. Some have tried to ensure access to product information by posting manuals in .pdf form along with videos and animations on their websites. Others have built instructional information right into the product itself. In addition to traditional word-based manuals, videos and web animations have become sources of product information. These visual media can be highly effective, but require just as careful planning and execution as a traditional manual.

Given the cost of translation, some manufacturers have begun to develop product documentation that relies exclusively on images rather than a combination of words and visual elements. While images do not need to be translated, they may require training to interpret correctly, and not every type of information lends itself to visual communication.

Usability testing is essential for all manuals to ensure that the information is conveyed effectively to the user. For manuals that rely primarily or exclusively on visuals to convey their message, usability testing is critical.

CHECKLIST: ALTERNATIVES AND SPECIALTY MANUALS

- Is my video or animation well-planned and scripted?
- Have I made sure the pace of my video is not too fast?
- Is everything shown in the video done exactly according to the instructions in the manual?
- Have I designed my video to be easy to update?
- Is my visual-based manual understandable without text explanations?
- Will the drawing conventions used be understood by my audience?

NOTES

1. The project seeks to provide ultra-low-cost, rugged laptops to children throughout the world to help further education. Each of these special laptops functions as a wireless router, and consumes so little power that it can be recharged by solar or human power. See http://one.laptop.org/about/mission for more information.
2. Weigant Merkt et al., "Learning with Videos vs. Learning with Print: The Role of Interactive Features," *Learning and Instruction* 21 (2011): 687–704.
3. Pew Research Center, March 2018, "Social Media Use in 2018." See also follow-up survey conducted in early 2019, "Share of U.S. Adults Using Social Media, Including Facebook, Is Mostly Unchanged Since 2018," www.pewresearch.org/fact-tank/2019/04/10/share-of-u-s-adults-using-social-media-including-facebook-is-mostly-unchanged-since-2018/, accessed April 29, 2019.

7 Special-Purpose Manuals

IS ONE ENOUGH? WHEN YOU NEED MULTIPLE MANUALS

Sometimes, manufacturers decide to put operator information into more than one manual. While it is common to have an operator manual and a separate service manual (see the next section), having multiple operator manuals is less common.

WHEN JUST ONE WON'T DO

Why would you need (or want) more than one manual? For most products, just one operator manual is preferable, but it is not always workable. The following are some of the reasons to consider multiple operator manuals:

- The product is composed of major systems produced by different manufacturers.
- The product has very different functions that may be performed by different people.
- The product is so complex that a single manual would be unwieldy.

Industrial machinery is often composed of large components built by different manufacturers. For example, the drive machinery may be made by one company, the conveyor by a different one, and the programmable controller by still another. The seller actually manufactures only the processing component of the machine. Some consumer products, particularly outdoor power machinery such as lawn mowers, pressure washers, or generators, may have a separate manual provided by the engine manufacturer, if it's different from the selling company.

Sometimes products have different functions that are performed by different people. For example, some industrial machines use complex programming that requires a specially trained technician, but once programmed, can be readily operated by others. Or a farm tractor with a power takeoff (PTO) may have a variety of attachments, each of which has its own separate manual.

Sometimes a product is complex enough that a single manual would be cumbersome and difficult to use. More commonly, a manufacturer decides to put additional information into the operator's manual beyond simply operation and routine maintenance. For example, an automobile manufacturer may include comprehensive information about driving under varied conditions, or a printing press manufacturer may want to provide detailed instruction on safety procedures. The resulting manual might be too big to be used with ease. One solution is to split the manual into separate books.

WHAT'S THE DOWNSIDE?

If it's difficult to get users to read one manual, it's doubly so to get them to read more than one. Before you decide to split information into multiple manuals, consider these questions:

- How can you allocate the information between manuals so that it makes sense and users will easily find what they need?
- Can you provide adequate safety information in each manual so that your user does not need to move back and forth from one book to the other to understand and avoid the hazards?
- If one book is lost, will the other still be usable without it?

Users quickly become frustrated if the organization of a single manual is confusing or makes it hard to find needed information. The frustration will only increase if they have to search through two or more books to find what they're after. As with single-volume manuals, the best organizational strategy is to arrange the content in terms of user questions. Put yourself in the user's shoes, so to speak, and allocate content in the way that makes the most sense from the user's point of view. If the user is looking for the recommended lubricant for roller bearings, would it make sense to put the disassembly procedure that gives access to the bearings in the same book? If the user is searching for how to sharpen the cutter blades, would it make sense to put the replacement procedure in the same book? Inevitably, you will have to make choices—otherwise, you'd be putting everything in one book—but make those choices on the basis of user convenience.

You can't be sure that manual users will follow a reference that sends them to another page in the same manual—for certain they will be even less likely to go to another manual entirely, and especially for a safety message. For that reason, if you plan more than one manual, make sure each one is self-contained in terms of safety messages. That does not mean you have to have every safety message twice if you have two manuals. Just be sure that every procedure described in a given manual has the associated safety messages in the same manual.

Closely related to the issue of avoiding making the reader jump back and forth from book to book is the question of whether one will be useful if the other is lost or destroyed. The more self-contained each manual is, the better. If one depends extensively on the other and you lose one, then you might as well throw the other away (unless, of course, you can download a replacement from the Internet!). Clearly, the most independent model is the product with manuals for separate components, like the pressure washer with one manual for the engine and another for the washer. The least independent is likely to be the one for the highly complex product that requires such extensive documentation that it simply won't fit in one book. Even in that case, by dividing subject matter functionally and putting together procedures that logically would be performed at the same time you will make the best of having to have more than one manual.

INSTALLATION, SERVICE, AND MAINTENANCE MANUALS

Up to this point, the focus has been on manuals that an end-user would read. Operator manuals are geared toward product users and focus (or should!) on how to use the product. Installation, maintenance, and service manuals are special-purpose books. Installation manuals are specifically geared to installing—not using the product. Service manuals typically cover repair and maintenance procedures for a single product, often a portable or mobile one. Examples of products for which you would expect to find a service manual include automobiles, pumps, sewing machines, and fax machines. Maintenance manuals typically apply to large, stationary, industrial machines: such things as milling machines, paper converters, injection molders, and turret lathes. These machines often have component parts (such as controllers or motors) that are manufactured by a different company than the one that built the machine. The maintenance manual must include information about maintaining those parts as well.

WHO ARE YOUR READERS?

While the principles of good verbal and visual design outlined elsewhere in this book will still apply (i.e., instructions should be presented in parallel structure, graphics should be clear and easy to read, and so on), the application of these principles will differ because of the difference in audience and purpose. Depending on the product, the audience for an operator manual can be anyone from a professional user who is technically sophisticated to a member of the general public who has never seen the product before, much less used it. By contrast, the audiences for installation, service, and maintenance manuals are presumed to have some technical knowledge and may be very familiar with the product. Just as with operator manuals, conducting some research to determine who your readers are will help you design well.

One important caveat is that installation manuals may have a considerably different audience from service and maintenance manuals. Some products, such as major capital equipment used in industrial settings, may be installed directly by the manufacturer, using factory labor. Others may be installed by the maintenance staff of the receiving industry. For example, an automobile manufacturer may use its own staff to install a new parts conveyor system. Either way, the installers are likely to be technically trained and familiar with the product type. Consumer products, on the other hand, may be installed by employees of the distributor (who may or may not be technically trained) or even by the consumer. Just as the audience for consumer product operator manuals can vary widely, so can the audience for consumer product installation manuals.

Some manufacturers try to avoid do-it-yourself installations by inserting language along the lines of, "This product should be installed only by qualified individuals . . ." The difficulty with that approach is twofold. First, it's unenforceable. Unless the manufacturer is prepared to refuse to sell products except to authorized dealers who use factory-trained installers, there is no way to prevent a purchaser from installing the product. Second, unless "qualified" is defined (e.g., "licensed electrician"), the statement is essentially meaningless.

What Goes Into a Special-Purpose Manual?

The purpose for an operator manual is to give clear instructions for a product's use and care. It introduces new users to the product and explains what it is for and how to make it work. An operator manual may give some simple maintenance procedures, such as how to clean the coils of a dehumidifier or how to change the oil in a lawn mower, but such instructions usually cover only the most basic operations. For anything complicated, the user is usually referred to "an authorized service representative." The service or maintenance manual, by contrast, is what the authorized service representative uses. The purpose of these manuals is to explain in detail the repair and maintenance of a product—to replace the compressor on the dehumidifier or overhaul the engine on the lawn mower. Normally, a service or maintenance manual will assume that the reader is familiar with the product, knows how to operate it properly, and needs specialized information only.

If you are assigned to develop and write an installation, service, or maintenance manual, what should you include? Do the same organizational principles work as for an operator manual?

Unlike an operator manual, an installation, service, or maintenance manual will usually not include any introduction to using the product. While it may peripherally address such issues as the product's capabilities or limits of operation, the focus is not on operating it, but rather on installing, maintaining, and repairing the product. Of course, the precise content will vary with the product, but certain categories of information are typically covered.

Installation Manuals

An installation manual will usually include the following:

- Location and site requirements
- Procedures for moving, unpacking, or uncrating the machine
- Installation procedures, including initial start-up and checking for proper operation
- Installation of special safety equipment

Installation manuals will usually provide recommendations for locating the product and explain detailed site requirements. These may include required clearances, floor strength specifications, procedures for reinforcing floors, and specifications for needed utility hookups, such as gas, compressed air, electrical power, or steam. Some products may have specific air-handling needs, including exhaust fans, fume hoods, dust collection, or other specialized ventilation requirements. Some products require special temperature or humidity controls. Whatever the specific site requirements are, the installation manual should provide the installer and new owner with detailed information.

Some products, especially very large capital machinery, have special requirements for lifting and moving the product. The installation manual may identify the location of lifting eyes or other lift points, specifications for placement or securing on a flatbed trailer, or other special requirements. The manual will usually include detailed instructions for unloading the machine and placing it on site. The manual may give procedures for unpacking or uncrating the product and removing packing material, including wood or plastic cribbing to prevent movement of parts in transit.

An installation manual should include step-by-step instructions for putting the product into service. A machine may need to be leveled, braced, or secured to the floor. Utilities may need to be connected. In addition to specifying what the utility needs are, the manual may include diagrams showing the proper locations for utilities. The manual usually will provide procedures for initial start-up and for checking to see that everything operates properly. (The first time that a product is operated, the procedure may be different from ongoing start-up.) Some products may have a diagnostic procedure or program that needs to be run on initial start-up to ensure that the machine is operable through its full range.

Some products, whether industrial or consumer, may have special safety equipment that should be installed when the product is put into service. For example, modern gas and electric ranges are built out of lighter materials than in the past. A new hazard that has emerged as a result is the possibility of the stove's tipping over (with potentially tragic results) if a small child stands on the open oven door. As a result, stove manufacturers provide anti-tipover devices. These are flanges that can be bolted to the floor. The stove is then slid into them, effectively locking the stove to the floor. The installation manual includes instructions for positioning and securing these devices. Special safety equipment that is put in place at installation often has no impact on the operation of the product, and thus would not be addressed in the operator's manual.

Service and Maintenance Manuals

A service or maintenance manual will generally address these topics:

- Specifications for the product, including capabilities, limits of operation, and capacities
- Recommended lubricants, cleaning agents, and other consumables
- Technical background on the function and operation of the product and its systems
- Operating tolerances and procedures for adjustment
- Routine maintenance procedures and recommended service intervals
- Troubleshooting guide
- Repair procedures
- Model variation information
- Parts catalog (may be a separate publication)
- Certain kinds of safety information

There is no set order for these elements. Some of these topics (especially in a service manual) may also be covered in an operator manual, although they will be treated differently.

One type of information that may not be included in an operator manual is the lockout/tagout procedure. For many industrial machines, maintenance procedures require that guards be removed to give access to parts of the mechanism. Obviously, the maintenance engineer or technician adjusting unguarded mechanisms may be at considerable risk if the machine is accidentally started. Some of these machines are very large, and an operator at one end may be totally unaware that someone is working on an unguarded mechanism at the other end. For that reason, a lockout/tagout procedure is commonly used. Such a procedure entails physically padlocking switches in the OFF position to prevent an accidental start. A special tag is placed on the machine to alert users that it is locked out.

Many companies have their own lockout/tagout procedures. It is nevertheless a good idea for the writer of a machine maintenance manual to include a recommended procedure in the manual. Information that should be covered includes

- The location of lockout points on the machine
- What each lockout point controls
- When the machine needs to be locked out
- Who should control the keys to the locks

Some companies put the maintenance supervisor in charge of the keys. This practice puts the burden on that supervisor to be utterly sure that it is safe to turn the machine back on. A safer, although potentially more cumbersome procedure is to have the individual worker lock out the portion of the machine that he or she is working on, and keep the key on his or her person until the work is done. That worker then puts the machine back in service. In any case, the manufacturer should provide guidance as to when and how a lockout/tagout procedure should be used. In addition to this general information, each procedure that requires the machine to be locked out must have a clear warning to that effect at the beginning.

Despite differences in content, many things are the same for all kinds of manuals. All of the organizational and writing strategies described in Chapter 3 apply to service manuals as well as to operator manuals. The use of such techniques as general-to-specific organization, lists, parallel structure, and active voice are just as important to the reader of the service manual. However, service and maintenance manuals differ from operator manuals in important ways that go beyond just the content. These include style differences and design differences.

STYLE DIFFERENCES

The differences in style between an operator manual and a service or maintenance manual have to do primarily with level of detail, level of language, pace, and tone rather than with the basic principles of presenting information.

Level of Detail

The information in a service or maintenance manual will naturally be far more complete than in an operator manual. More complete explanations will be given, and more complex procedures will be described. Service manuals will contain detailed information on specifications and tolerances. Operator manuals will contain some of this same information, but only what is necessary for routine maintenance by the operator. For example, the operator manual for a portable weather radio would include specifications for power input from various sources (batteries, AC current, car battery), with specifications for appropriate adapters. A garden tractor owner's manual might include specifications for a spark plug gap, on the assumption that the owner might do his or her own tune-ups. A service manual, however, would be much more detailed. The increased detail typically means that the text contains more numerical information.

Dealing with numerical information such as specifications or tolerances in a primarily textual context presents problems. When you include specification information in the text of technical background and procedure sections, you must be especially careful to avoid the following:

- Letting the numbers get lost in the paragraphs of text
- Letting the numbers obscure the flow of your description of how an assembly works or how a mechanism should be adjusted

This requires constant attention to how numbers relate to the rest of the text. If you have one or two numbers in a long paragraph of explanatory text, the reader can all too easily skim over them and miss what may be vital information. On the other hand, a paragraph loaded with numbers is terribly hard to read. Generally, more than four or five exact numbers in a paragraph of text is too many. The reader simply cannot keep the numbers straight and often loses the line of thought conveyed in the text.

Chapter 5 recommends how visuals could be used to make numbers easier to follow. Some options for handling occasional numbers within text also work. To make an occasional numeral stand out in a sea of words, print it in boldface. As an alternative, the numbers may be separated by white space from the surrounding text. If the number occurs in a procedure description, try to put the exact number in a step of its own rather than including it with other adjustments. Example 7.1 shows how this technique can clarify the text.

Example 7.1: Using Steps to Separate Numbers from Text

Original

Position the fan motor in the fan housing so that there is 2–3 mm clearance on all sides of the fan and secure with the 4 adjustable clamps. If adjustment is needed, tighten or loosen the clamps one at a time, no more than ¼ turn each time, until the proper clearance is achieved.

Improved

1. Position the fan motor in the fan housing, securing with the four adjustable clamps, maintaining 2–3 mm clearance on all sides of the fan.
2. If adjustment is needed, tighten or loosen one clamp at a time, turning it no more than ¼ turn each time.

If the text contains many exact numbers, it may be better to put them in a separate table or chart. A set of exact numbers can be much more easily assimilated in chart form than in paragraph form. The table need not be elaborate or formal. It can consist of simply setting the numerical information off from the rest of the text by white space. Example 7.2 shows how this technique improves readability.

Example 7.2: Use of an Informal Table to Separate Numbers from Text

Original

Whenthe new extractor is installed, check the fit as follows: Insert a caliper and measure the distance between the hook of the extractor and the opposite side of the breech face. The measurement must be between 7.2 mm and 7.3 mm for the Model 72 and between 6.9 mm and 7.0 mm for the Model 75. If the space is greater than tolerance, file the adjustment pad on the extractor until it is within tolerance.

Revised

When the new extractor is installed, check the fit. Insert a caliper and measure the distance between the hook of the extractor and the opposite side of the breech face. The measurement must be within these ranges:

Model 72 7.2 mm—7.3 mm
Model 75 6.9 mm—7.0 mm

If the space is greater than tolerance, file the adjustment pad on the extractor until it is within tolerance.

The last technique for making numerical information visible is to include tolerance and specifications in visuals accompanying the text. In designing your visuals, you must be careful that numerical information does not clutter up the drawing. A good drawing can easily be ruined with too many labels and excessive specification information, especially in a service manual. Since the audience is generally more technically sophisticated than the audience for an operator manual, the writer may be tempted to use unedited engineering drawings for visuals. Although such drawings contain a wealth of specification information, they are usually too cluttered to be useful for service or maintenance procedures. It is certainly possible to use drawings well to convey specification information, as Figure 7.1 and Figure 7.2 show. Note especially the use of the close-up to illustrate a particular portion of the drawing. Be sure, if you use drawings to present specification or tolerance information, that the drawing is placed next to the relevant text, particularly in procedures sections.

Language Level

The language in a service manual will be a good deal more technical than the language in an operator manual. Since anyone reading a service manual (presumably) has a certain amount of technical expertise, the writer can use a more specialized vocabulary. Don't make things technical just for the sake of making them technical, however; good writing of any sort is as simple as it can be and still convey the necessary information precisely.

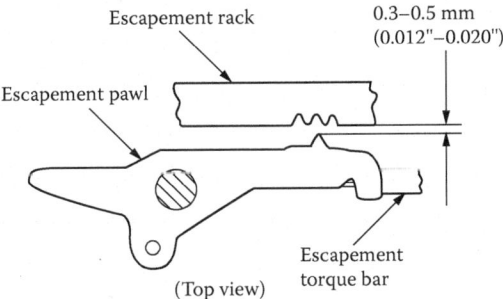

FIGURE 7.1   Example of a graphic used to convey specifications and tolerances.

Source: From Single-Element Typewriter, Model 200, Service Manual, Silver Reed America, Inc., Torrance, CA, p. 44. With permission.

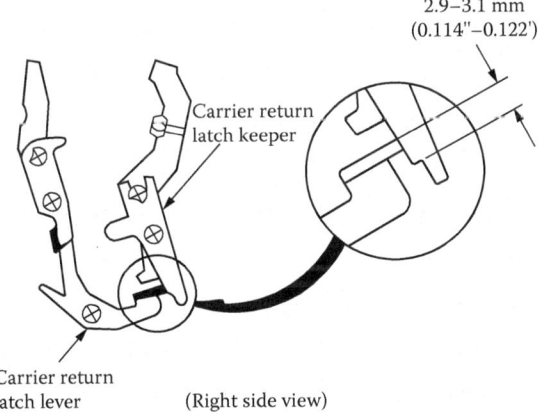

FIGURE 7.2   Example of a graphic that conveys information about tolerances. Note the effective use of a close-up to avoid cluttering the drawing and to make it more readable.

Source: From *Single-Element Typewriter, Model 200, Service Manual,* Silver Reed America, Inc., Torrance, CA, p. 45. With permission.

Remember also that not all readers of the manual will be familiar with the manufacturer's nomenclature for parts. If you use a term that is special to one manufacturer, try also to include the generic name for the item as well—particularly if you are referring to a tool. It is common practice for factory-produced service manuals to refer to tools by factory numbers instead of generic names (e.g., "VW 558" instead of "flywheel holding fixture"). The writer of a service manual must be aware that not all users of the service manual will be authorized, factory-trained technicians. It might appear that using a specialized terminology for specialized tools would discourage amateurs; however, it is more likely that the amateur will simply find some other way to do the procedure (e.g., using a screwdriver rather than a snap-ring pliers to remove a snap ring). If a specialized tool is necessary for a given procedure—for safety or to avoid damage to the product—note this fact and refer to the tool by name as well as by number.

Pace

The pace of a service manual may also be somewhat faster than that of an operator manual. This simply means that you can present information at a faster rate on the page. You may include less background information and more substantive words per sentence. Remember, however, that the emphasis in a service or maintenance manual is on procedures, which means that the reader will probably be looking back and forth between the manual and the product as he or she performs a procedure. Keep your sentences and paragraphs relatively short, and use formatting devices to make it easy for the technician to find the right place in the manual again after looking away for a moment to do a step in a procedure. Overloaded sentences and paragraphs are just as annoying to a technically sophisticated reader as they are to a first-time operator.

Tone

Finally, the tone of a service manual may be less conversational than that of an operator manual. As we have noted, the function of an operator manual is in part to represent the company to its customers and in part to gently introduce the new user to the product. This requires that an operator manual be written in everyday language and that it sound "friendly." The purpose of a service manual is primarily to explain to a professional or knowledgeable amateur technician how to perform various repair and maintenance procedures. Example 7.3 shows the differences in tone. The example contains two excerpts, the first from a typewriter owner's manual and the second from a typewriter service manual, both dealing with the operation of the right margin stop. The technology may seem dated (although you can still buy a brand-new typewriter on Amazon!), but the principle is as valid as ever.

Example 7.3: Style Differences between Operator and Service Manuals

Operator Manual

The right margin stop prevents you from typing past the right margin; however, you can space or tab right through it. To type past the right margin, press MAR REL (margin release) and continue typing.

Service Manual

As the Carrier moves to the right still more after the Bell ringing, the Margin Stop Latch moves up the Margin Stop Right extension allowing the Margin Rack to rotate. Then the Margin Plate attached to the Margin Rack rotates the Linelock Bellcrank through the Margin Link. At that time, the Linelock Bellcrank extension moves the Linelock Keylever downwards causing its extension to insert into the space between the Keyboard Lock Balls. And the Keyboard has been locked to prevent typing past the Margin Stop Right.

Sources: Operator manual reprinted from *IBM Correcting Selectric II Operating Instructions*, International Business Machines Corp., White Plains, NY, © 1973, p. 6. Used with permission. Service manual reprinted from *Single-Element Typewriter, Model 200, Service Manual*, Silver Reed America, Torrance, CA, p. 33. Used with permission.

DESIGN DIFFERENCES

As with operator manuals, the needs of the user determine the appropriate choices for design elements in service and maintenance manuals. Users of service and maintenance manuals are concerned with maintaining and repairing products. Unlike operators, they will not ordinarily ever read through the manual to get a sense of the product. Instead, they will always be looking for specific information on how to perform specific procedures. The design of a service or maintenance manual must make it easy to use. Generally, these aspects of manual design may be handled differently in these more technical manuals:

- Organization
- Page size and layout
- Binding system
- Graphics

Organization

Service and maintenance manuals are often organized by product system; for example, an auto service manual will have chapters on the engine, the transmission, the cooling system, the electrical system, and so on. This kind of organization makes it easy to find the needed information. If a car has a problem with the cooling system, the technician knows to go to that chapter first. Frequently, individual chapters will have their own tables of contents.

This organizational system works very well when the technician or maintenance engineer knows the precise location of a problem. However, the systems of any product interact, and the cause of the malfunction may not be immediately obvious. Even if the user goes to the proper section, other information located elsewhere in the manual may be important as well. Because the manual divided up into sections does not easily show the overlap of systems, the writer must take care to include a comprehensive index and cross-referencing as appropriate. If possible, include cross-referencing within the index; for example, "fuel injectors, cleaning (see also fuel filter)." The idea is simply to make the manual useful to the reader by making it as easy as possible for him or her to find the section needed.

The problem of how to organize the information in a service manual becomes particularly difficult when the manufacturer decides to combine the service manual and the operator manual in one book. Although one book is cheaper to produce than two and ensures that everyone has the same information, using one manual for both is not the best practice. The service manual has a very different audience and purpose than the operator manual, and this difference should be reflected in content, style, and format. To combine both kinds of manuals in one book makes it nearly impossible to maintain the appropriate distinctions. Furthermore, including the service manual with the operator manual may encourage some users to perform procedures they should not perform because they aren't skilled enough or don't have the appropriate tools. It is much better for the manufacturer to keep operator manuals and service/maintenance manuals separate.

Page Size and Layout

Almost all service and maintenance manuals use 8½ × 11-inch pages. These manuals are usually used in a shop or factory, and kept in the technician's or engineer's office. They do not need to be as "portable" as operator manuals, and the larger page size allows for more readable illustrations. Because the 8½ × 11-inch size has become so standard, it's best not to deviate from it. Odd-sized manuals are hard to file and easy to lose.

The large page size dictates, to some extent, the layout. Certainly, a single-column layout running the full width of the page would be difficult to read. Some technical manuals use a two-column format, but the 2/5 layout is more common. This gives a column width of approximately five inches—still a very readable length for 10- or 12-point type. In addition, the remainder of the page provides space for illustrations and notes. It's a good idea to include some blank pages at the end of each chapter for additional notes.

As with operator manuals, the typeface should be chosen for easy reading. A font size of 10- to 12-point is ideal. It should not be smaller than 10-point. Even though these manuals are usually used indoors, the lighting may be poor.

Binding System

Service manuals may be bound (perfect or spiral bindings are common), but machine maintenance manuals are usually placed in three-ring binders. This loose-leaf format has several advantages:

- New information, such as technical bulletins, can be easily added.
- Corrections or revisions can be sent to users in the form of replacement pages, saving having to reprint the whole manual.
- Related information, such as manuals for related or component equipment, can easily be inserted into the same binder.

Whatever sort of binding is used, loose-leaf or otherwise, it must be able to withstand hard use. Where an operator manual may be read once and then filed away to lie untouched unless a problem occurs, a service or maintenance manual will be consulted regularly. Even if a technician or engineer is thoroughly familiar with the product, he or she will need to check specifications and tolerances. Both the cover and the pages will need to be sturdy enough to withstand frequent use in potentially dirty environments. A plastic or plastic-coated cover is usually a must for a service or maintenance manual, as is somewhat heavier-than-normal stock for the pages, particularly if the manual is in a loose-leaf format.

Graphics

As with the verbal portion of a service or maintenance manual, the basic principles for visual design (explained in Chapter 5) apply to both operator manuals and service/maintenance manuals. Good visual design is perhaps even more important in the more technical manuals because so much of a service or maintenance manual is devoted to procedures for repair and adjustment. The combination of verbal and visual material must make the procedure perfectly clear. This often means that the balance of material shifts toward the visual: a service or maintenance manual will tend to have more drawings and photographs than an operator manual. Although the principles are the same for both kinds of manuals, again their application differs. A service or maintenance manual will have more technical drawings, exploded or assembly diagrams and cutaways rather than perspective drawings, and circuit diagrams rather than block diagrams. You must take great care to ensure that these are large enough to see easily and are not cluttered.

An exploded diagram, for example, can often be made much more comprehensible by dividing it into sections. Figure 7.3 shows how "sectionalizing" an exploded

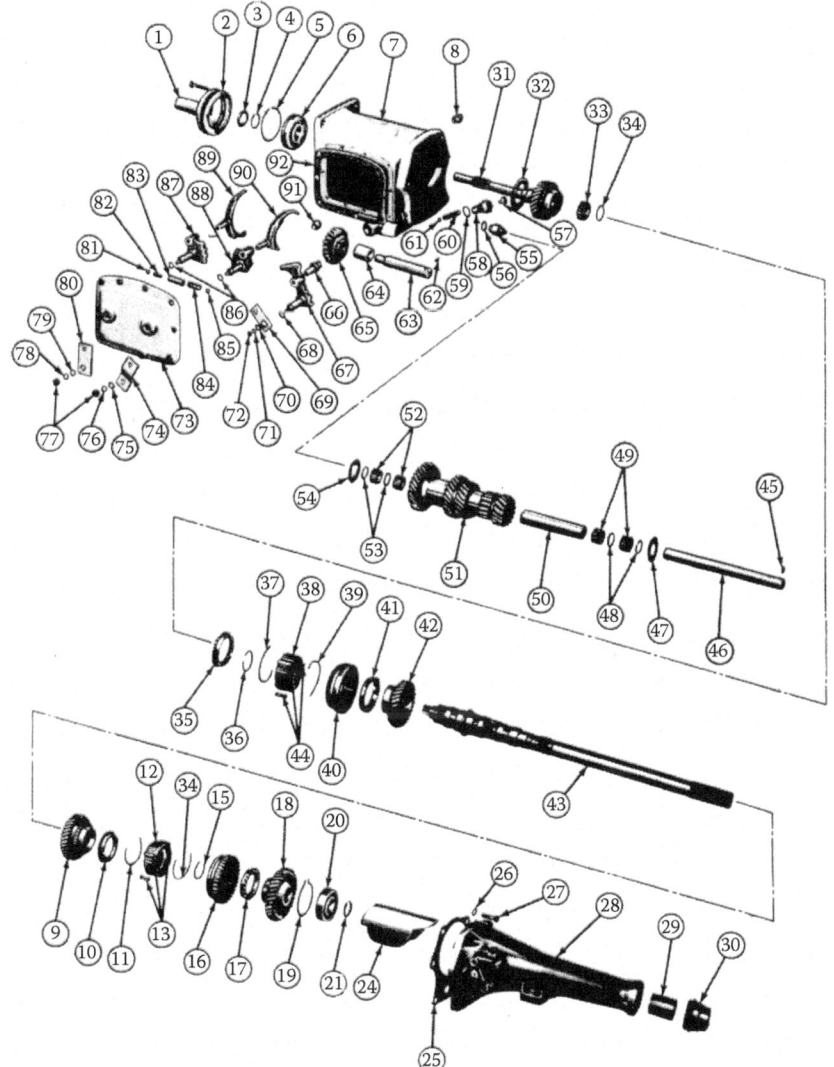

FIGURE 7.3   Example of "sectionalizing" an exploded view to make it appear less cluttered and to enable the reader to view smaller parts in close-up.

Source: From *Dodge Dart, Coronet, and Charger Service Manual*, Chrysler Motor Corp., Dodge Division, Detroit, MI, 1967, p. 21–18. With permission.

drawing of a transmission permits more complicated portions to appear in close-up. The whole view could have been laid out in one piece and photo-reduced to fit on the page, but the result would have been difficult to read.

As suggested in Chapter 5, when the complexity of the drawing permits, label parts with part name rather than a callout. Since the technician or engineer using the manual will already be looking back and forth between the manual and the product, adding another place to look (the key that identifies the callout) will only increase the possibility of a mistake. You must also ensure that the lines showing how parts fit together are easily distinguished from lines or arrows leading from labels or callouts. Typically, this is shown by using broken lines for the former and solid, heavier lines for the latter.

If you use cutaway drawings, be sure that the reader can easily differentiate the "layers" of the cutaway. Often you can do this by careful shading—but be sure that your shading does not clutter the drawings. At other times, the best choice may be to use color to highlight different levels.

Perhaps the most easily abused form of illustration is the circuit diagram. The writer of a service or maintenance manual should avoid the temptation to use a reduced version of the schematic developed with the product. First of all, the original schematic was probably designed on a CAD station to be plotted on a large scale. Reducing it to fit onto the manual page would render it unreadable. Second, particularly for a service manual, it may contain more detail than the technician needs, which can lead to unnecessary clutter. Instead, have a schematic drawn or edited for the manual, one that includes only necessary information and is scaled appropriately for the manual page size.

If your product's manual includes a circuit diagram, you must be sure that your readers can interpret the symbols used. For a device that is primarily electronic, this is not usually a problem. Someone without knowledge of electrical circuitry is not likely to use a service manual for a portable two-way radio, for instance. However, for a product in which an electrical system is only one component—farm equipment, for example—and for which the service manual's users are likely to be diverse, you may wish to include additional information, such as a table of electrical symbols used and their meanings.

For machine maintenance manuals, you may need to include design drawings or complicated schematics. Rather than reduce these to fit an 8½ × 11-inch page, one option is to make the drawing the original size, but to put it in the manual as a fold-out. The easiest way is to place the folded drawing (label visible) in a plastic envelope that is designed to fit in a three-ring binder. Alternatively, the folded drawing can itself be punched to fit into the binding. While punching the drawing to bind directly into the manual may seem to increase the likelihood that the drawing will stay with the manual, the reality is that it is likely to be taken out of the binder for easy viewing, making it all too easy to tear out the holes. A plastic envelope is a better choice.

Good visual design simply means making sure that your visuals are big enough to be easily seen, as simple as they can be while still conveying the necessary information, and clear enough to be easily understood. How these basic principles are put into practice depends on your audience. Since the audience for a service or maintenance manual is likely to be more knowledgeable than the audience for an operator manual, you can use more technically sophisticated visuals—but you should still make them big, simple, and clear.

KEEPING SERVICE AND MAINTENANCE MANUALS CURRENT

As Chapter 8 discusses, the manufacturer has a continuing duty to warn of newly discovered hazards or changes in recommended procedures, providing that the manufacturer can reasonably be expected to reach those users. Because service manuals and maintenance manuals are, for the most part, used by professionals rather than the general public, it is easier for a manufacturer to locate them and provide updates. Certainly, factory-authorized service technicians will maintain a relationship with the manufacturer, and expect to receive periodic technical bulletins and other updates. Similarly, manufacturers of custom industrial machinery typically have an ongoing relationship with their customers, making it possible to provide them with new information.

Service manuals and maintenance manuals, unlike operator manuals, are normally updated periodically. The product user does not need update information, since he or she is unlikely to buy successive versions of the same product. In cases where that does happen—a company, for example, may replace office machines on a regular basis—the manufacturer usually supplies a new operator manual with each product.

This approach is impractical, however, for the service technician who repairs and maintains a manufacturer's product line over a period of time. One technician, for example, may service several generations of a given manufacturer's small gasoline engines—as well as those of other manufacturers. Especially when year-to-year modifications are relatively minor, it is much cheaper and handier for the manufacturer to supply supplements to an existing service manual than to write a whole new manual each time the product changes slightly. The writer of a service manual must keep in mind this need for frequent updates and must design the manual so that updating is easily accomplished.

Similar problems appear when a company's products are predominately custom installations, as may be the case, for example, with industrial packaging equipment or medical equipment. Somehow, the company still has to provide an accurate service manual. Digital technology is making this easier, particularly with a modular manual: one can, with the right system, custom-assemble a manual for each machine.

Many different techniques are used to update manuals. Some manufacturers supply replacement or supplemental pages to be inserted into the service manual, which, of course, requires that the service manual be bound in a loose-leaf binder. Increasingly, manufacturers are using video and other digital technology to keep their service technicians informed of changes and new procedures. For example, an auto manufacturer may supply its dealerships with video showing a new service procedure being performed, rather than requiring the dealership to send a mechanic to the factory for training. New information may be posted on the manufacturer's website or emailed to registered repair technicians. Although all these technological improvements in communication are desirable, the fact remains that not all users of the service manual or maintenance manual necessarily have access to the technology to use them. Even if such technology is available, it's wise also to provide "hard copy" versions—including supplements to the manual. These supplements are usually in the form of individual pages or separate pamphlets.

The former method—replacement or add-on pages—is better for two reasons:

- If new information is bound right into the original manual, the manufacturer is assured that the technician has the new information. Separate booklets are too easily lost or misfiled.
- If the new information is bound into the manual at the relevant place, it is more likely to be noticed.

To ensure that the update pages are used, the writer of a service manual can use the following techniques:

- At the beginning of the manual, tell the reader that supplements will be provided from time to time and explain how to use them. Distinguish between add-on pages, which should follow existing pages in the book, and replacement pages, which require removal of the old pages.
- Provide a log at the beginning or end of the book on which the technician can record the addition of supplemental material.
- Number pages clearly according to the original manual's pagination and build in a means of distinguishing add-on from replacement pages. (For example, add-on pages might be numbered with the page number of the page they should follow, plus a letter—26a, 26b, etc.—whereas replacement pages would simply be numbered the same as the pages they are intended to replace.)

You may also include in a service manual supplemental information about probable field modifications. This information tells the service technician of ways in which the owner may have modified the product. Although these modifications may not be approved by the manufacturer, they do take place, and the technician needs to know of them. On the other hand, if a particular modification appears often, it may be a signal that a design change is needed in the product. If a field modification produces a hazard, such as the removal of guards or bypassing of interlocks, technicians should be directed to reverse the change and put the owner on notice that the modification is not acceptable. Sales representatives and local dealers are good sources of information for the service manual writer about what modifications may be expected.

Manuals as Training Texts

Service manuals often serve as "textbooks" for training technicians. In factory training programs and in technical schools, students learn by doing. Large manufacturers often conduct their own training programs for certified repair technicians. Product service and maintenance manuals, supplemented by other instructional materials such as videos, workbooks, etc., are typically used as training manuals. Technical colleges train technicians to repair all kinds of mechanical, hydraulic, and electronic products. An auto repair student in a transmissions class at a technical college, for example, will learn how to fix transmissions by working on one—from a specific make and model of car. The service manual for that car will be the primary resource for the student learning how to do various procedures. In fact, some manufacturers partner with technical colleges to do their "factory-authorized" training for them.

If you know that a service or maintenance manual will be used for training, what, if anything, should you do differently when you are writing and designing the manual? If you know that your manual will be used as a training tool, you may wish to specify what sections in the manual must be included in a training program. Some manufacturers go so far as to develop formal training curricula that include an outline for the training program, references to specific manual chapters, and even test questions. If your manual is used often for training, a relatively formal system of certifying personnel makes sense.

In any case, make certain that the procedures in the manual reflect the best practices. Do not recommend or even describe "shortcuts" that could be hazardous. Keep the information in the manual current. Make sure that any photos and illustrations of persons performing procedures reflect proper practices (personal protective equipment in place, machine locked out, proper tools used, etc.). The text and the illustrations must not conflict. Remember that images are generally more powerful and memorable than words.

MANUALS FOR CUSTOM MACHINERY

One of the challenges for writers of manuals is the custom machine. Unlike consumer products, which typically may have a few different models that are then sold to thousands (or even millions) of consumers, large capital machines are often custom designed for a specific purchaser's application. The net result is that a company may sell many different "models," but in effect have only one customer for each model. How to manage creating manuals (operator manuals as well as installation, service, and maintenance manuals) for these one-of-a-kind products is a challenge. The challenge is compounded when the machine uses components made by another manufacturer.

THE ONE-TIME MANUAL

Creating a custom manual is a good deal easier when your "custom" machine differs from others in limited ways. In that situation, you can easily assemble most of the manual (whether operator or service) from modules that apply to all the products in a particular line and write the remaining sections that are unique to that specific machine. It is more difficult if your products really are custom designed to the buyer's specifications. Even then, you will probably find that some aspects of the machine are common to more than one model, so you will not have to write an entire manual for each machine.

One of the most difficult aspects to solve satisfactorily is the illustrations in the manual for a custom machine. If you use generic illustrations, they may not adequately reflect the specifics of a given machine. Developing custom illustrations will drive up the cost of the product documentation. One possible solution is to adapt design drawings for use in the manual, particularly if it is a service/maintenance manual. All the caveats apply: be sure that the drawings clearly show the desired elements, make certain that the lettering on callouts or other labels is large enough to read easily, and make sure that the reproduction is good so that detail is not obscured by excessive reduction.

Producing a one-of-a-kind manual is challenging and usually will be more expensive than a standard manual, but the cost is also more readily built into the price and passed on to the customer.

Managing Manuals for Component Parts

Many large capital machines, whether custom designed or not, employ component parts created by other manufacturers. The original equipment manufacturer (OEM), for example, a company that sells fire engines, may use a Ford chassis and engine with an Allison transmission, Waterous fire pumps, a Hale compressed-air foam (CAF) system, and a Hannay high-pressure hose reel. Who writes the manual and what should it include?

In a situation such as this, you have two choices: try to tie everything together and write a comprehensive manual for everything or give the customer a fistful of separate manuals for each component. OEMs handle the problem in different ways. One of the more common options is for the OEM to develop and write a comprehensive operator manual, but pass on the individual service and maintenance manuals provided by each component manufacturer. The idea is that the operator needs to use the product (whether it is a fire engine or a box folder) as an integrated unit, so the operator manual should make that as easy as possible. The fire engine manual would include instructions for setting the transmission in the proper configuration before starting the pumps, instructions for operating the CAF system, and so on. On the other hand, anyone performing service or maintenance procedures will do so on individual systems. Just because you need to replace a leaky seal in a fire pump, you would not necessarily need to repair the hose reel. For that reason, having individual service manuals for component parts works just fine.

SUMMARY

Although the digital revolution has changed how we do business, it has not turned us into a paperless society. We still need manuals, and in fact, sometimes we need more than one. When your product requires multiple manuals, ensure that each manual is as self-contained as possible, especially with respect to safety information. Users should not have to go to a different book to get the information they need to be safe. You may need to repeat certain information to be sure each manual user has what he or she needs, but the alternative is potentially much more costly.

Special-purpose manuals, such as installation, service, and maintenance manuals, usually have a much more limited—and specialized—audience than operator manuals. This more restricted audience results in differences in content, style, and even overall design. Whereas an installation manual may be used once and discarded, a service or maintenance manual will be referred to repeatedly over a long period of time. That long, useful life for service and maintenance manuals creates a need for ways to keep them updated with new information, particularly safety-related information. Using technological means such as online updates is one end of the spectrum; sending out replacement pages to put into a three-ring binder is the other end. Either way, good practice requires that technicians have access to current information, particularly when it is related to safety. Safety is especially critical when the manual is used as a training text, as repair manuals often are.

Original equipment manufacturers, whose products contain component parts made by other manufacturers, have a special challenge: how to make a manual for a "composite" product. The challenge becomes even greater when the product is not only made of component parts, but is also a one-of-a-kind custom installation. As always, keeping the focus on the user and the user's needs will yield the best result.

CHECKLIST: SPECIAL-PURPOSE MANUALS

- Have I included complete installation instructions, including instructions for unpacking?
- Have I included all needed specifications, including tolerances, capacities, and limits of operation?
- Have I included schematics for all systems?
- Is there a complete troubleshooting/diagnostic guide?
- Have I used generic names for tools or parts as well as proprietary ones (or included a glossary)?
- Does the page design make it easy to use? Is there enough white space? A place to jot down notes?
- Is there a system in place for providing updates?
- Is there complete safety information, including warnings and lockout/tagout procedures?
- Is the binding sturdy and dirt-resistant? Will the pages stand up to hard use?
- Have I included all OEM manuals for component parts?

Part III

You Have Been Warned

8 The Duty to Warn

OVERVIEW

One of the most difficult aspects of manufacturing and marketing products is ensuring product safety and minimizing the potential for liability. Products liability law, particularly in the United States where it is most developed, is a very complex area of law. It also continues to evolve, which makes it something of a moving target. Manufacturers are understandably worried about whether their products will pass muster with respect to product safety.

This aspect of the marketplace also elicits strong opinions on whether products liability law is—to borrow a phrase from *Fox News*—"fair and balanced." You will not find an answer to that question in this book. What you will find is an introduction to what the current state of the law suggests about what your approach should be to address product safety and to make choices about warning labels and safety messages in collateral materials such as manuals. I am not an attorney and do not practice law. You will not find definitive answers as to exactly what warnings you should or should not put on your product.*

What you will find is generally agreed-upon principles for how best to manage your product's hazards and risks and information about U.S. and international standards for hazard communication. Understanding these principles requires a little background in the development of products liability law in the United States. Of necessity, this background will be general and will neglect many of the twists and turns of legal history. Think of it as a "Poet's Physics" class in U.S. products liability law.

PRODUCTS LIABILITY LAW: WHAT YOU DON'T KNOW *CAN* HURT YOU

U.S. law is divided into two types: criminal law and civil law. Criminal law (naturally) covers crimes such as murder, burglary, or forgery. In a criminal law case, the defendant is alleged to have violated a statute that explicitly identifies what are called the elements of the crime. Under Wisconsin law, for example, if a person is to be convicted of the crime of theft, the prosecutor must prove that the defendant did four things:

- Knowingly
- Took the property of another
- Without consent
- With intent to deprive the owner of its use indefinitely

---

* Even if I *were* an attorney, I couldn't offer definitive answers. As a lawyer friend of mine puts it, "Two lawyers can't agree to kill a rat in a bathtub" (Valerie Van Brocklin, personal conversation).

Whether the defendant did those things may be difficult to prove, but it all really comes down to yes/no questions. In criminal law, the state brings the action against the defendant, and the state's attorney is a prosecutor.

Civil law, on the other hand, deals with disputes between parties that may or may not be based upon statutory or regulatory provisions. The parties are often arguing over the interpretation of rather general statutes that often are full of words like "reasonable." A judge or jury must decide, for example, whether damage to an apartment constitutes "normal wear and tear" or whether the landlord is entitled to compensation. In civil law, one party (the plaintiff) brings suit against another (the defendant). Each side typically hires legal counsel. There is no prosecutor—the plaintiff and defense attorneys present their cases and the judge or jury decides in favor of one or the other. The losing side may have to pay money to the winning side, but no one goes to prison as a result of the decision in a civil case.

The burden of proof is different in criminal and civil cases as well. In criminal cases, the normal burden of proof is "beyond a reasonable doubt." That is the most stringent test. In civil law, the burden of proof may be "preponderance of the evidence" or "clear and convincing evidence" depending on the type of case.

Because civil statutes are more general and more subject to interpretation than are criminal statutes, judges' and juries' decisions in similar cases may carry the weight of precedent. In other words, if the same situation happened two years ago in the same state, the current case is likely to be decided the same way. Over time, a body of case law builds up. Attorneys involved in civil cases will cite prior decisions to bolster their side of the argument. But because civil law tends to offer general principles that are subject to interpretation rather than yes/no elements to prove, there's a lot of room for argument, and over time, the decisions evolve. While a current case might be decided just as a similar case from two years ago, the decision might be very different from that in a similar case from twenty years ago.

WHEN IS THE SELLER LIABLE?

In general terms, for a plaintiff to win a products liability suit, he or she has to prove four things:

- The product was defective in some way.
- The defect was present when the product left the seller's control.
- Injury or damage occurred.
- The injury or damage was caused by the defect.

The plaintiff must prove all four. If any one of these is absent, the seller is not liable. To take a simple example, suppose you buy an electric drill at a hardware store. Electric drills are generally double-insulated, meaning that even if the motor develops a short circuit, the outside casing will not become energized and shock the user. Your new drill is defective and that protection is absent. As you are taking your new drill out of the box, you drop it on your foot and break your toe. Can you sue the manufacturer? Certainly—but you won't win. While the product is defective, and the defect was probably present when it left the factory, and you are injured, the defect and the injury are unrelated. On the other hand, if you dropped the drill on your toe because you received an electric shock, you might have a case after all.

Note that the heading asks, "When is the seller liable?" and not "When is the manufacturer liable?" As the lawyers put it, liability attaches throughout the chain of distribution. If you are injured by a defective drill, you can sue the hardware store where you bought it, the distributer who sold it to the store, and the manufacturer who made it. That fact may become a concern if your company uses component parts to make its product.

LEGAL THEORIES

Most products liability lawsuits are brought under one or more of the following three legal theories:

- Breach of warranty
- Negligence
- Strict liability in tort

The earliest products liability lawsuits were brought under breach of warranty, but today, most at least involve a claim under strict liability. Here's what the three theories mean.

Breach of Warranty

Very simply put, a claim under breach of warranty says that the product did not perform as the seller said it would. The earliest lawsuits were brought against sellers of food, on the idea that if you are presenting food for sale to the public, you are warranting that it is safe to eat. The difficulty with breach of warranty is that usually only the purchaser has standing to sue, because the warranty extends only to the purchaser. Warranties may be express or implied (public sale of food would be an example of an implied warranty).

Negligence

Negligence means that the seller did not exercise reasonable care in the design, manufacturing, or marketing of the product. For example, the maker of a meat slicer that has an unguarded blade located between the operator and the ON/OFF switch could be found negligent because the design is unreasonably dangerous. Or the manufacturer of electrical wire with too much variation in the thickness of the rubber insulation on the outside of the wire could be found negligent because the manufacturing and quality control processes did not ensure reasonably uniform insulation.

Strict Liability in Tort

Torts are civil wrongs (as opposed to criminal) that are not based on contracts. The plaintiff need not have any sort of contractual relationship with the seller. Strict liability means the seller is liable if the product has a defect rendering it not reasonably safe, even if the seller exercised reasonable care—in other words, even if the seller was not negligent. An example might be a flawed glass baking dish that cracked and broke under ordinary oven temperatures, even though the manufacturer's quality control procedures met industry standards. Strict liability concentrates on the product, not on the care with which the manufacturer operated. Strict liability is the predominant legal theory applied in cases involving warnings.

It's Not Fair—Or Is It?

At first, this concept seems unfair. How is it that a manufacturer or distributor can be held liable when it did nothing "wrong?" And how can manufacturers and other sellers protect themselves against lawsuits? And what's reasonable?

The rationale for strict liability is twofold. First, it is seen as an incentive (albeit a negative one) for manufacturers to design safe products and to provide adequate instructions and warnings. Second, it is seen as a way of sharing the cost both of injury and of safety. Because strict liability focuses on the product, not the manufacturer, it encourages manufacturers and sellers to insure themselves against liability claims. Buying insurance increases the cost of doing business, and those increased costs are passed along to the buyer in the form of higher prices. Thus, a single manufacturer or consumer does not bear the whole cost of an injury caused by an unsafe product, and manufacturers and consumers who benefit from safer products also pay part of the cost of increased safety.

What's reasonable? As a University of Wisconsin Law School professor was fond of saying, "Go to court, and the judge will tell you." Reasonable means you don't have to be perfect. It means you do not have to take measures to keep people safe that would be so costly that they would render your product non-competitive in the marketplace. But it also means that if there are measures you could take that would increase safety and would not put an undue burden on you, then you should take those measures. The "reasonable person" test essentially asks, "What would a reasonable person in a similar situation do?" Very often, that is precisely the question that the judge or jury must answer. As we shall see, the law recognizes that finding that balance between safety and efficiency is not easy.

While strict liability may seem unfair from a manufacturer's perspective, it has resulted in demonstrably safer products—but it has also no doubt resulted in some potentially very useful products never coming to market because of safety concerns. Fair or not, strict liability is the growing trend in U.S. law.

Geographical Variations—United States and Abroad

As noted, products liability law in the United States is built upon case law. Some states do have products liability statutes, but like most laws governing non-criminal matters, these are general and subject to interpretation. Other states rely on the general principles of tort law and the rules of commerce to cover products liability disputes. Because states vary and case law varies, the same set of facts may produce different results in different states. Products liability cases can also be brought in federal court, in which procedures and rules may be quite different from those in state court.

Are you confused yet? Products liability law is indeed confusing and constantly changing. It is difficult even for attorneys who specialize in that area of practice to keep up. Non-specialists—and especially non-attorneys—cannot possibly do so. Fortunately, in areas of the law that depend heavily on evolving case law, from time to time leading legal scholars will publish assessments of the current state of the law in that area. These are called "restatements." The most recent products liability restatement is the *Restatement of the Law Third, Torts: Products Liability.*[1] It is an excellent resource on products liability law (although challenging to read) and has sections devoted specifically to the issues surrounding instructions and warnings. One caution is in order. Not all states have adopted all of the provisions described in the *Restatement Third.* You should consult with your company's legal counsel for guidance on your state's view. Nevertheless, the specific parts that relate to instructions and warnings are widely accepted.

In the United States, in part because of the reliance on case law to inform questions of liability and in part because of increasing reliance on the legal theory of strict liability, a product that meets current established standards could still be found to be unreasonably dangerous. Failure to meet regulatory requirements and standards may be *prima facie* evidence that a product is defective, but the reverse is not true. Meeting standards is considered the minimum that a manufacturer must do. Note also that not all products have associated standards to guide their design and manufacture.

The situation in the European Union and much of the rest of the world is different. Internationally, if your product meets established regulations and standards, it is generally presumed to be reasonably safe, and the injured plaintiff has a greater burden to prove that the product is dangerous. The European Union has included safety standards as an integral part of its overall product regulations—the Machinery Directive, for example, requires that only safe products be placed in commerce. Other countries have passed specific laws addressing product liability. An example is Japan's Product Liability law.[2] While not all countries have specific laws or standards relating to products liability, for U.S. manufacturers, the good news is this: if your product would be considered reasonably safe in the United States, it almost certainly would also be considered reasonably safe in the rest of the world.

FAILURE TO WARN IN PRODUCTS LIABILITY LITIGATION

As noted in Chapter 1, plaintiffs are increasingly alleging "failure to warn" as a cause of injury.[3] Under strict liability, any product defect can form the basis for a product liability claim, assuming the plaintiff can show that the defect caused the injury. It may be easier to convince a jury that a defective or absent warning caused an injury than that a defect in the product itself caused the harm. To take a slightly far-fetched example, let's consider the electric drill you dropped on your toe a few pages back. The drill was defective—it was not properly double-insulated—but that defect did not cause your injury, so a products liability lawsuit based on that defect should be a non-starter. On the other hand, an attorney could argue that the manufacturer should have warned that the drill was heavy and could cause injury if dropped. Such a lawsuit would have a chance of success, either in the form of an offer of money to settle the lawsuit or a favorable jury verdict if the case went to trial, because inadequate warnings or instructions can be considered a product defect.

While we tend to think of the defect as implying that something about the product is malfunctioning, like a brand-new lightbulb with a broken filament, that kind of malfunction is not the only kind of defect as the word is used in the context of products liability. A product defect may be a design defect, a manufacturing defect, a packaging defect, a marketing defect, and so on. A product defect is anything that renders the product not reasonably safe.

Poor instructions and warnings can indeed render an otherwise acceptable (i.e., in terms of design, manufacturing, etc.) product defective. According to the *Restatement Third,* instructions and warnings are inadequate and the product is defective when "the foreseeable risks of harm posed by the product could have been reduced or avoided by the provision of reasonable instructions or warnings . . . and the omission of the instructions or warnings renders the product not reasonably safe."[4]

The trend in products liability litigation of more emphasis on failure to warn is likely to continue for several reasons, including these:

- Safer designs
- More automation
- Expectation of safety
- Perceived ease of warning

Safer Designs

Today's products are much safer than similar products just a few years ago. In part because of regulatory action and litigation resulting in companies placing greater emphasis on safe designs, it's much more difficult now to prove a design to be defective. Table saws have guards and anti-kickback pawls, garage doors have photoelectric beams and reversing pressure switches, industrial meat slicers have electric interlocks, and glass-forming machines have light curtains that shut down the machine if the operator comes too close to moving parts. Automobiles have backup cameras and collision alarms. But no product can be completely safe, so warnings are still required, and often require judgments about content and placement.

More Automation

As products become more automated, they also become more complex. Manufacturers must decide how much information about how these automated systems work should be included in the user manual. In many cases, the automated systems are designed to increase both efficiency and safety. For example, a modern automobile may automatically apply the brakes if it senses that a forward collision is imminent. As systems become more complex, and especially as products interact with other products independent of human input through the Internet of Things, it becomes more difficult to provide the user with comprehensive information about product operation.

A news story current as of this writing illustrates this dilemma perfectly. Boeing's 737 MAX aircraft was involved in two fatal crashes, one in Indonesia in October 2018 and one in Ethiopia in March 2019. Preliminary investigation suggests that the cause of both crashes may have been a malfunction in the plane's automatic anti-stall system that prevents the nose of the airplane from rising too high, leading to an engine stall. A malfunctioning sensor may have signaled that a stall was imminent when in fact, the pitch of the aircraft was normal. The automated system then responded by pushing the nose down, and even though the pilots tried repeatedly to counteract the system, they were ultimately unsuccessful and the planes crashed.

One of the issues that has emerged is that the manuals available to the pilots and the training given them did not fully address how the automated system operated and how to override it.[5] Whether or not the anti-stall system was in fact the cause of the crashes, the challenge for manufacturers remains the same: as products become more automated—often to increase safety—how much information about automated systems should be included in the operator manual and what warnings are needed regarding possible malfunctions of automated systems? The problem will be compounded as the Internet of Things results in more and more products being connected—and influenced—in ways the manufacturer cannot easily foresee.

EXPECTATION OF SAFETY (OR, IF I GET HURT, IT MUST BE SOMEBODY'S FAULT)

Without a doubt, products are safer today than they used to be, which is a good thing. At the same time, increased product safety has also driven the misguided perception that if someone gets hurt using a product, the product must somehow be defective. Products do not have to be *absolutely* safe, only reasonably safe. Further, the manufacturer is not required to warn of every conceivable hazard, nor must every warning be perfect. How do you figure out what's reasonable? Even the authors of the *Restatement Third* acknowledge the difficulty, but they also give guidance as to what to consider: "No easy guideline exists for courts to adopt in assessing the adequacy of product warnings and instructions. In making their assessments, courts must focus on various factors, such as content and comprehensibility, intensity of expression, and the characteristics of expected user groups."[6] Despite its difficulties, products liability law is not entirely capricious.

PERCEIVED EASE OF WARNING

Nevertheless, it's easy for a plaintiff's attorney to assert that an additional warning would have prevented the client's injury. It seems like a simple and inexpensive proposition to add a warning that covers the circumstance of the injury. Whether it's adding an additional label on the product or an additional warning in the manual or both, it *seems* to be an easy fix, but for the manufacturer it is more complicated. It's neither possible nor advisable to try to warn about every imaginable hazard. How do you decide when a warning is needed and what it should say?

Part of the answer comes from the hazard analysis discussed in Chapter 2. Warnings should be reserved primarily for residual hazards, those that cannot be eliminated in another way. But do all residual hazards need a warning? If not, how do you decide? The next section addresses those questions.

WARNINGS—THEY'RE NOT ALL WACKY
(BUT SOME SURE ARE!)

- WARNING! Do not attempt to iron clothes while wearing them.
- WARNING! Do not use microwave to dry pets.
- WARNING! Remove baby from stroller before folding for storage.

In 1997, an organization called Michigan Lawsuit Abuse Watch (M-LAW) began sponsoring an annual contest to select the top "wacky warnings" submitted to it—and there has been no shortage of submissions to choose from. Few topics will generate more discussion, confusion, opinions, and frustration among manufacturers than that of safety warnings. We have all heard stories of people who received multimillion-dollar products liability awards after suffering injury doing something patently stupid—like picking up a rotary lawn mower and trying to use it as a hedge-trimmer, or attempting to dry a wet poodle in a microwave oven. Technical writers and editors are both anxious for information on how to write good warnings and defensive about the need to do so: "You can't idiot-proof the product" is a common sentiment. The good news is that you don't have to. The cases alluded to here actually never happened. They were described in an insurance industry advertisement explaining rate hikes.*

While there are from time to time multimillion-dollar court awards, the reality is much more prosaic. The vast majority of lawsuits—about 95%—are settled out of court, but even when they proceed to trial, the average money payout (from a settlement or jury award) in 2017 was less than $53,000.[7] Nevertheless, even a $50,000 settlement can be a significant burden for a small company—or result in higher products liability premiums. And even if you win a case or settle it successfully, it costs money to defend. While it is impossible to lawsuit-proof your product, you can minimize your chances of being sued by knowing when and how to warn.

When Do You Need a Warning?

In general, the manufacturer has a duty to warn of hazards present in the product unless the hazards are "open and obvious." The manufacturer also has a duty to warn of hazards that would result from "foreseeable misuse." Let's take a look at what each of these terms means.

Open and Obvious

There is no duty to warn of hazards that are "open and obvious" to users of the product. For example, you don't need to warn that a knife has a sharp blade that can cut. That is considered open and obvious because it is part and parcel of the purpose of a knife. Similarly, there is no need to warn that when you light charcoal briquettes, they get very hot—that's what they're designed to do, and if they didn't get hot, they would not cook food. On the other hand, there may well be a duty to warn that when you light charcoal briquettes, they give off deadly carbon monoxide gas (a normal by-product of combustion), so they should only be used outdoors where the gas can harmlessly dissipate.

---

* I routinely offer $100 to the first seminar attendee who locates a valid case citation for either of these. I will extend the offer to readers of this book.

To take another example in which the hazard, though open and obvious, is not directly related to the purpose of the instrument, consider an ordinary hammer. Everyone who has used a hammer to pound nails (its intended purpose) has probably also inadvertently struck a thumb or finger. While that is not a hammer's intended use, it still is an open and obvious hazard that requires no warning. However, hammers typically do carry a warning to wear safety glasses—not because you are likely to smack yourself in the eye, but because of a hazard that is neither immediately obvious nor widely known. Ordinary hammers are made of hardened steel, and common nails are not. When used to pound a 10-penny nail into wood, the use of the hammer poses no special risk (although if a nail is just barely penetrating the wood it is possible to hit it off-center and cause the nail to pop out of the wood). The real danger comes if the hammer is used to hit something else made of hardened steel—doing so can cause small metal fragments to fly off at high speed, posing a significant risk of eye injury. That hazard is not "open and obvious" and so a warning is in order.

Even if the danger is open and obvious, the manufacturer may have a duty to warn if the user might not be aware of the extent or degree of danger. For example, a person using tile adhesive labeled "flammable" probably would not smoke while using the product, but might well not realize the danger posed by pilot lights, especially those in a different room. *Russell v. Mississippi Valley Gas*[8] dealt with a person who received severe burns when he was cleaning paint brushes with gasoline in a storage room attached to a garage. The pilot light of a nearby water heater ignited the fumes.

Be careful about too-easily dismissing a hazard as "open and obvious" and therefore needing no warning. What is obvious to you, working in the industry, may not be so obvious to someone else. For example, people working in the tire industry know that if you replace only two tires on your car, the new tires should always be put on the rear wheels, regardless of whether the car has front-wheel drive or rear-wheel drive. Most of us not working in the tire industry would expect that the new tires should go on the drive wheels. But putting new tires on the front wheels creates a hazard when driving on wet pavement, because the better tread on the front may contribute to oversteering—the front tires grip the pavement, but the rear ones hydroplane, potentially allowing the rear end to slide around and overtake the front, leading to loss of control. Just as "shop-blindness" can cause you to omit needed information when writing instructions, it can also cause you to overestimate your users' knowledge of product hazards.

Foreseeable Misuse

Manufacturers also have a duty to warn against foreseeable "misuse" of the product—a term that is somewhat misleading. Essentially, misuse simply means a departure from the instructions for safe use. The term "misuse" does not imply a reckless disregard for safety. The key word here is "foreseeable." The manufacturer is not required to cover every imaginable misuse of a product. Liability attaches only "when the product is put to a use that is reasonable to expect a seller or distributor to foresee. Product sellers and distributors are not required to foresee and take precautions against every conceivable mode of use and abuse to which their products might be put."[9] Thus, a manufacturer of chlorine-containing laundry bleach certainly has a duty to warn of potential skin irritation, since this is not an obvious risk inherent in normal use of the product to bleach clothes. However, the manufacturer may also have a duty to warn against mixing the product with ammonia (as someone might do if using the bleach as a household cleaner), which produces deadly chlorine gas. Mixing bleach with an ammonia-containing cleaner is a reasonably foreseeable, although not intended, use. Similarly, a manufacturer of glass casserole dishes may have a duty to warn that a dish taken out of a hot oven and placed on a cold surface may break (reasonably foreseeable), but would not have a duty to warn against breakage caused by using the same casserole dish to drive nails (not reasonably foreseeable). Again, "reasonableness" is the test.

If an unforeseeable misuse actually happens—somebody tries to pound nails with a casserole dish—does that mean that particular misuse has just become foreseeable and must be warned against? Not necessarily—a single "off-the-wall" misuse of a product does not automatically mean more warnings. One consideration is the frequency of occurrence. Suppose one person decided to use a glass casserole dish as a hammer and was injured when the dish shattered. How many glass casserole dishes are in circulation? If the occurrence is extremely rare (not to mention unlikely), probably no warning would be required. On the other hand, if a particular misuse shows up repeatedly, a warning is probably in order. You will still have to make decisions about the value of adding a warning versus the "dilution effect" that additional warning will have on other warnings.

WHO NEEDS TO BE WARNED?

The manufacturer has a duty to warn anyone who might reasonably come into contact with the product. Thus, the manufacturer may be held liable for injury to someone other than the person who actually bought the product. For example, the buyer of industrial chemicals may be a manager or purchasing agent, but the actual users will be workers on the floor. The buyer of a fleet of golf carts may be the owner or manager of a golf course, but the users will be golfing customers. This requirement that warnings reach the actual user of a product may mean that warnings need to be located on the product itself as well as in accompanying literature, because the user may not have access to the manual.

The situation for the technical writer is further complicated by the fact that the potential users of a product may comprise a very diverse group. You must consider the possible range in terms of age, sex, expertise, familiarity with product, even literacy. In *Hubbard-Hall Chemical Co. v. Silverman*,[10] the court held that a written warning was not adequate because it failed to provide for illiterate users. More commonly (at least in the United States), the issue is not illiteracy *per se*—rather, it is inability to read English. In the United States, the fastest-growing minority group is Hispanics, some of whom speak only Spanish. In general, recent immigrants from any country who have minimal English skills will tend to be concentrated in lower-paying jobs, many of which involve handling dangerous chemicals (pesticides, industrial solvents) or potentially hazardous machinery. Warnings need to be useful to them. Consider providing the warning in Spanish as well as English.

Sometimes the problem is that those coming into contact with the product are not intended users at all. For example, what kind of warning would reach children with access to typical (but toxic) household chemicals? Or what sort of label should go on a high-voltage transformer in an area near a playground? These two problems have resulted in creative solutions. The standard symbol in the United States denoting poison has for years been the skull and crossbones (see Figure 8.1). In fact, Environmental Protection Agency (EPA) regulations require that symbol on pesticides.[11] In Pittsburgh, the skull and crossbones has a very different meaning: it's associated with the Pittsburgh Pirates baseball team, so the Pittsburgh Poison Center at Children's Hospital in Pittsburgh developed the "Mr. Yuk" symbol (see Figure 8.2) to warn children of toxic substances. In the case of electrical transformers, the lightning bolt that indicates high voltage also is associated with the super hero The Flash. So, in 1981, the National Electric Manufacturers' Association (NEMA) developed the "Mr. Ouch" symbol shown in Figure 8.3. Like Mr. Yuk, Mr. Ouch is copyrighted, but is available for a small licensing fee.

FIGURE 8.1 Skull-and-crossbones symbol for poison. The meaning of this widely used symbols is not intuitively obvious, but must be learned. For that reason, it may not be suitable for all users.

Source: From 49 Code of Federal Regulations, chapter 1, § 172.430.

FIGURE 8.2 "Mr. Yuk" symbol for poison. This symbol was developed specifically to communicate to children, using an intuitive image suggesting that the labeled substance tastes bad.

Source: Copyright © Pittsburgh Poison Center. With permission.

FIGURE 8.3   "Mr. Ouch" symbol for dangerous voltage. This symbol was developed specifically to communicate to children, using an intuitive image suggesting a painful shock.

Source: Reprinted from NEMA 260–1996 Safety Labels for Pad-Mounted Switchgear and Transformers Sited in Public Areas, by permission of the National Electrical Manufacturer's Association.

As you consider warning labels for your products and safety messages for use in manuals and other materials, be sure to consider not only your intended users, but any others who might be likely to come in contact with the product.

CAN YOU HAVE TOO MANY WARNINGS?

In short, yes. Not every hazard requires a warning. If the hazard is open and obvious, generally no warning is required. If the hazard is not known at the time of manufacture, and is not foreseeable, no warning is needed—although you may have a duty to warn if a hazard becomes apparent after the sale.[12] But even if a warning is not required, wouldn't it be a good idea to include all that you could think of? Is there any downside to overwarning?

Actually, there is a downside. Warnings have two purposes: to inform users about hazards they might not know about so they can avoid them, and to remind users of hazards they're already aware of, but might become complacent about. People are already skeptical about warnings, especially in light of the silly ones out there. The more warnings you put on a product or include in a manual, the less attention each individual one will receive. Users' eyes glaze over and they dismiss them all—and miss the critical ones. As the *Restatement Third* puts it, " . . . warnings that deal with obvious or generally known risks may be ignored by users and consumers and may diminish the significance of warnings about non-obvious, not-generally-known risks."[13] In fact, the research shows that this is precisely what happens.[14]

Most manufacturers put too many warnings on their products and in their manuals in an attempt to protect themselves against every imaginable products liability claim. The result is that they fail to reach the user about the most important hazards, either because the critical warnings are lost in a sea of messages or because the user sees one silly warning and ignores the rest. I once consulted for a company that made gravel separators used in quarries. The original label on the product listed (in rather small print) 30-some separate warnings, one of which was, "Do not ingest hot hydraulic oil." When asked, company officials admitted that they'd never actually heard of people drinking hydraulic oil (hot or cold), but it probably wouldn't be good for them if they did!

A much better strategy would be let your comprehensive and well-documented hazard analysis (see Chapter 2) guide you: to develop on-product safety labels only for the most serious and likely residual hazards. The remainder can be addressed in the manual.

KEEP THE USER SAFE: COMPREHENSIVE SAFETY INFORMATION

On-product labels are one avenue to warn users, but by no means the only avenue. Ideally, all aspects of the materials accompanying products—labels, packaging, manuals, instructional videos, and promotional materials should work together to help keep the product user safe.

ON-PRODUCT SAFETY LABELS

Your hazard analysis identified and ranked hazards in terms of severity and likelihood. On-product labels should address the most severe and likely residual hazards—those that cannot be designed out or guarded. If reaching in to clear a material jam in a rotating auger can result in an arm being caught and amputated in the blink of an eye, then an on-product warning is needed. In the United States, the most familiar format for on-product safety labels is that detailed in the American National Standards Institute (ANSI) Z535.4 standard *Product Safety Signs and Labels*. This is one of a series of standards addressing communication of safety information. Chapter 9 goes into considerable detail on the design of on-product labels according to both this standard and its international counterpart, which is ISO 3864–1. In this chapter we will using three terms relating to the ANSI Z535 standards:

- Signal word
- Safety-alert symbol
- Signal-word panel

Signal Word

ANSI uses three injury-related signal words: DANGER, WARNING, and CAUTION, and a fourth, *NOTICE*,[15] for important information not related to personal injury. These words represent varied levels of hazard severity and likelihood as defined in the standard:[16]

- DANGER: Indicates a hazardous situation that, if not avoided, will result in death or serious injury. This signal word is to be limited to the most extreme situations.
- WARNING: Indicates a hazardous situation that, if not avoided, could result in death or serious injury.
- CAUTION: Indicates a hazardous situation that, if not avoided, could result in minor or moderate injury.
- *NOTICE*: Indicates information considered important, but not hazard-related (e.g., messages relating to property damage). The safety-alert symbol shall not be used with this signal word.

 This is the safety alert symbol. It is used to alert you to potential personal injury hazards. Obey all safety messages that follow this symbol to avoid possible injury or death.

FIGURE 8.4   The safety-alert symbol. This symbol, used with and without signals words, conveys the potential for a personal-injury hazard.

Source: Reprinted from ANSI Z535.6–2006 American National Standard for Product Safety Information in Product Manuals, Instructions, and Other Collateral Materials by permission of the National Electrical Manufacturer's Association.

Note that the only difference between DANGER and WARNING is the likelihood of injury, *not* the severity. In other words, DANGER is reserved for situations in which it is *almost certain* that a person interacting with the hazard will be killed or sustain serious injury. For example, in the printing industry, a common unsafe work practice is for a worker to use a rag folded into a small pad to clean excess ink from the printing press rollers while the press is running. The hazard is that the moving rollers form an ingoing nip. The operator assumes that if the rag gets caught, he can simply let go of it. Unfortunately, a person's reaction time is too slow for that. By the time the worker's brain recognizes that the rag is caught and sends a signal to the hand to release it, the worker's arm has already been drawn into the nip. What is the appropriate signal word? In this case, WARNING is the proper choice. Printers persist in this unsafe practice because most of the time wiping the moving roller does not result in injury, because the rag doesn't get caught, so the injury (crushed arm), while severe, is not *almost certain* to happen.

Safety-Alert Symbol

The safety-alert symbol is an exclamation point inside of an equilateral triangle, as shown in Figure 8.4. It simply indicates an injury hazard, without specifying the nature of the hazard.

Signal-Word Panel

A signal-word panel consists of the safety-alert symbol plus one of the three signal words that indicate injury hazard (DANGER, WARNING, CAUTION), surrounded by a box.

SAFETY MESSAGES IN MANUALS AND PACKAGING

We tend to think of the manual as providing information about how to use and maintain the product—and so it should. But the manual is often overlooked as a safety resource as well. Sure, the attorneys demand the "safety page" in the front of the manual, and perhaps a few warnings sprinkled here and there in the text, but that's about it. In fact, it's not uncommon for companies to assign the tasks of developing the manual and labels to totally different work groups. Failing to coordinate the two can cause real problems—and undercut an opportunity to use the manual to leverage the impact of the labels. Similarly, decisions about packaging are often left to the marketing department because, after all, the package is designed to sell the product, right? Yes—but it does much more.

On-product labels are a critical means to deliver essential safety information to the user, but they are necessarily limited in what they can do. So, what do you do with other important safety-related information? As a manufacturer, you have two primary venues for safety information and several secondary ones. The primary ones are the manual that accompanies the product and the packaging that surrounds the product. Each has strengths and weaknesses as a means to deliver safety information.

The manual gives you the space to include not just warnings, but also a fuller explanation of each hazard and how to avoid it. It is likely (but not guaranteed!) to be kept by the user. It allows multiple opportunities to warn of the same hazard encountered in different operations. On the other hand, the manual, though retained, may not be read by the user.

The packaging, unlike the manual, will certainly be seen by the user in most cases. You can be assured that a warning placed on the packaging will reach the user—or at least the installer. On the other hand, you can also guarantee that in most cases the packaging will immediately be discarded. So, while you have a sterling opportunity to present a warning that will initially reach the user, that warning probably has a very short effective life.

COORDINATING MANUAL AND LABELS

How do you decide what to put where? Ideally, you want your on-product labels and all the collateral materials that go with the product—manual, quick-start guide, packaging, and advertising to present a consistent picture. Each piece is a little different, but they all should work together to form a coherent whole. Chapter 11 describes an approach called Integrated Product Safety that seeks to build safety into every aspect of product development, but here are a few guidelines specifically for safety information in the manual:

1. If there's a label on the product, put a corresponding warning in the manual.
2. Separate safety messages that address general hazards and safe work practices from those that address hazards specific to your product.
3. Use the label to refer the user to the manual.

If There's a Label on the Product, Put One in the Manual

Any hazard significant enough to warrant an on-product label certainly warrants treatment in the manual as well. Do the two warnings have to be exactly the same? No—in fact, they probably should not be identical. However, some things should match:

- Signal word
- Avoidance information

Both safety messages should use the same signal word (e.g., "DANGER" or "WARNING"). The criteria for choosing a signal word have to do with the severity of the hazard and the likelihood of injury should the user interact with the hazard. Those factors don't change whether the warning is on the product or in the manual.

Both warnings should have the same avoidance information—don't have the label direct the user to "keep hands away from" moving parts when the manual tells the user to "exercise caution when working near moving parts." How can you be "away from" and "near" at the same time? However, while the avoidance information must not conflict, it does not have to be identical. The manual may go into more detail. On-product labels are extremely limited in space, so the word message must be kept very short and to the point. Manuals have more flexibility, so you can elaborate. While the label may merely say, "Keep hands away" from moving machine parts, the manual can explain how far away is enough and how to use the product safely without getting hands too close to moving parts. While the label may say "Lockout/ tagout machine before performing maintenance," the manual can include a recommended procedure.

Be sure you do not give a mixed message. For example, if an on-product label warns to lockout/tagout the power before performing any maintenance, be sure that all the procedures in the maintenance section include locking out/tagging out the machine as a first step. If some maintenance needs to be performed with power on—so that the machine can be "jogged," for example—then rethink the warning to lockout/tagout the power before performing *any* maintenance.

Separate General Safety Warnings From Those Specific to the Product

When you pick up a manual for a product and start to read the safety section, you will often find that many of the messages say things like, "Keep electrical power cords dry," or "Keep work area free from debris." These messages address general safe work practices, and they may be appropriate to include in the manual. However, many manuals contain more or less the same messages, and as a consequence, people may tend to skip over them, thinking that they have read them many times before. If warnings that are specific to your product are mixed in with these generic messages, users may overlook them.

One way to address this problem is to group your safety messages into logical categories: electrical hazards, flammable vapor hazards, etc. You may also want to have a category called "safe work practices." Taking this approach has three advantages. First, it is easier for users to process information that is separated into smaller chunks. An undifferentiated list of 30 safety messages is unlikely to be read—after the first few, most users will ignore the rest. Second, by dividing them up and giving them headings, you are assisting the reader to understand the information—they already know the general topic before reading the first warning. Third, you can make it clear which warnings apply specifically to your product and which are simply reminders of good work practice.

Use the Label to Refer the User to the Manual

One of the difficulties for manual writers is that you can never be sure that the user has access to the manual. That means that the only safety resource the user has available is the on-product labeling, right? Not necessarily. If you use the on-product labels to refer the user to the manual, the user now knows that more information is available (somewhere) and may be able to access it directly or via the Internet. It is good practice to add "Read instructions before use" or a similar phrase at the end of the word message on a warning label. That way, the user is put on notice that the warning label may not tell the whole story and you will not feel as compelled to cover the product with labels. Some manufacturers print a QR code right on the on-product warning label that takes the reader to the relevant section of the manual.

Does safety information need to be presented in any particular way in the manual? Until a few years ago, the answer was no—and the variability from one manual to the next was remarkable. The publication of ANSI Z535.6, *for Product Safety Information in Product Manuals, Instructions, and Other Collateral Materials* offered new guidance.[17] This standard uses the safety-alert symbol and the same set of signal words as ANSI Z535.4 specifies for on-product labels, but provides flexible formatting options for safety-related messages in manuals. Chapter 9 explains in more detail on how these two standards work together to provide the user with complete and consistent safety information.

OTHER AVENUES FOR SAFETY MESSAGES

Secondary avenues for safety information include package inserts, "quick-start guides," follow-up marketing materials, videos, and other collateral materials. Don't underestimate the value of the Internet in getting safety information to your users. Post your product manuals on your Web site in .pdf form for download and your users will always have access to the manual—whether or not they can remember where they put the one that came with the product. Use your company website to highlight safety features of your products and to alert users to hazards or unsafe practices that weren't anticipated when the manual was written. Provide streaming video or animation to demonstrate proper product operation or maintenance procedures.

MANAGING THE POST-SALE DUTY TO WARN

Because the focus in strict liability is on the product rather than on the conduct of the manufacturer or seller, sellers have a continuing duty to warn of hazards that are discovered after the time of sale, even if you no longer manufacture that product. This duty depends on certain conditions being met. The rule is simple: "One engaged in the business of selling or otherwise distributing products is subject to liability for harm to persons or property caused by the seller's failure to provide a warning after the time of sale or distribution of a product if a reasonable person in the seller's position would provide such a warning."[18] To meet the reasonableness test, four conditions must be met:[19] [17]:

1. The seller knows (or should know) of the risk.
2. The users can be identified and don't know the risk.
3. A warning can be effectively conveyed and heeded.
4. The risk is great enough to warrant putting the burden to warn on the seller.

Companies often worry a great deal about the post-sale duty to warn. In practice, it is not invoked all that often, because the four conditions listed above are not all that frequently met. Typically, manufacturers have four questions:

- If I improve safety, does that mean earlier models of my product are defective?
- If I discover a hazard, should I recall the product or provide a retrofit solution?
- How can I find product users, once my product is sold?
- What kinds of safety upgrades should I offer?

Let's look at each of these.

DOES IMPROVED SAFETY MEAN EARLIER MODELS ARE DEFECTIVE?

In short, the answer is no. The idea of a post-sale duty to warn is a relatively new concept, and the courts have not universally embraced it, particularly if the product was not defective at the time of sale. An example might be a home desk manufactured in the late 1950s with a pull-out shelf intended for a portable typewriter. It turns out that, while it is of adequate strength to support a small typewriter, the shelf can collapse under the weight of a desktop computer and monitor. That anyone would have a computer for home use was inconceivable in 1950—prior to the invention of the printed circuit and the silicon chip, a computer would fill an entire room. The desk was not defective at time of sale, because the use of it to support a personal computer was not foreseeable.

Post-sale duty to warn generally involves hazards that were not known at the time of original sale. The fact that safer designs have since been developed does not mean that older designs are defective. While the manufacturer may have a continuing duty to warn of newly discovered hazards (providing the users can be found), it does not automatically have a duty to inform users of each safety improvement that has been made since the original purchase. As the *Restatement Third* puts it, "If every post-sale improvement in a product design were to give rise to a duty to warn users of the risks of continuing to use the existing design, the burden on product sellers would be unacceptably great."[20] Such a duty would also serve as a disincentive to improve safety.

SHOULD I RECALL OR RETROFIT?

Some recalls are not voluntary. Products that are regulated by the federal government, such as medicines and food, may be subject to forced recalls. The Consumer Products Safety Commission requires injuries from consumer products to be reported, and that information may prompt a recall. Other recalls are voluntary, when a company decides that it is better to bear the cost of recalling a product and replacing it free than to risk the liability exposure from leaving it on the market.

Whether a retrofit of safety equipment or other safety upgrade is a feasible alternative depends on the product, on the defect, and on the ability of the user to install the retrofit. For consumer products, for example, the recall is more common simply because the manufacturer can publicize the recall notice and rely on retail distributors to install the fix for the problem, whereas it would be difficult, if not impossible, to identify individual consumers to send them a part. One exception, of course, is automobiles, for which owner information is reported at time of sale. The manufacturer can communicate directly with the owner—or at least the original purchaser—to give instructions or provide safety equipment.*

How Can I Find Product Users?

This is by far the thorniest question of all, especially when the danger is discovered long after manufacture and sale of the product, as is often the case with prescription drugs or other medical products. Despite such products going through testing, sometimes unexpected side effects can take a long time to appear. A similar issue arises with industrial machinery. This sort of product typically has a very long useful life. A machine may be used by the original purchaser, then resold, and sold again. Even more difficult are consumer products, for which there may be no record at all even of the original purchaser. With prescription drugs and some medical products, you have a physician's prescription to start with. For large capital machinery, the manufacturer will usually have a record of the original purchaser at least, and that provides a place to start. But with consumer products, you have none of that.

One option is to use media, especially targeted media, to alert users to newly discovered safety issues and give them information for addressing them. If, for example, you sell tire-balancing machines and have discovered a problem in a particular model, you could buy advertising or place an article in various trade journals aimed at tire shops, service stations, and other businesses that deal with tires. Another is to approach the problem indirectly. For example, suppose your company manufactures powder-actuated tools, commonly known as "stud guns." These are widely used in the construction industry, and it would be difficult to find all the users. On the other hand, anyone who uses a powder-actuated tool on a regular basis needs to buy the powder charges that operate it. Working with the powder-charge manufacturer to get the word out may reach nearly all the current users of the product. As you think about how to reach users, broaden your scope to include required maintenance or resupply needs as avenues to locate purchasers.

What Kind of Safety Upgrades Should I Offer?

There is no standard answer for this question. The choice you make depends on the hazard and how easily it can be fixed. For many hazards, simply providing additional or upgraded labels may be enough. For others, replacement parts or add-on guards, etc., might be in order. What you need to do is conduct a "mini" hazard analysis of the newly discovered hazard. Go through the same process as you would during product development of figuring out everything you can about the nature and severity of the hazard, when it occurs, what the conditions of use and user characteristics are, and so forth. Then follow the same hierarchy of responses.

---

* Sometimes these are extreme. One recall notice from an auto company, for a faulty owner's manual, instructed owners to return the car to the dealership so a technician could "install a corrected manual in the glove box."

Usually, designing the hazard out is not feasible on an already-existing product. But providing a way to lock out the machine may be almost as good. Perhaps you can retrofit a guard for a mechanical hazard. If you become aware of a procedure that is performed in an unsafe way, you can provide alternative instructions. For example, suppose the manual for your panel-cutting machine directs the user to remove the blade for sharpening and use a specially designed sharpening tool. You have learned that the tool often gets lost and the blade is cumbersome to remove, so users tend instead to reach in and simply hold a sharpening stone against the rotating blade, posing a serious risk of injury if their hand slips. You might identify these possibilities for how to manage the problem:

- Retrofit a guard that uses an electronic interlock to prevent access to the blade unless the machine is locked out.
- Design a sharpening tool that can be used from a safe distance (a sort of stone-on-a-stick, for example) and that can be tethered to the machine.
- Design a warning label to be placed near the hazard directing users to lock out the machine and remove the blades for sharpening.

Which alternative—or combination of alternatives—you choose should be based on the same sort of cost-benefit analysis that you would use in a standard hazard analysis.

One final recommendation: do not require users to pay for after-the-fact safety items, whether they are new warning labels or retrofitted guards. They won't buy them. Instead provide them at no cost. The fact is, if your company is concerned enough about a post-sale safety issue to go to the trouble of trying to fix it, your liability exposure is likely to be potentially a lot more expensive than providing the fix for free.

SUMMARY

Products liability law is complex and ever-evolving, making it a challenge for anyone to keep up, let alone a technical writer. However, it is based on fundamental principles and standard legal theories, including breach of warranty, negligence, and strict liability in tort. Increasingly, plaintiffs are relying on strict liability in tort as the basis for their claims. This legal theory focuses on the product and whether it is defective rather than on the seller's conduct. Product defects can include inadequate instructions and warnings; as designs improve, the legal focus turns more to manuals and the warnings that accompany products. Not every hazard requires a warning— and too many warnings can be nearly as bad as too few—but trying to decide how to manage potential hazards can be a daunting task.

The best approach is to use a comprehensive hazard analysis to guide your decisions about what to warn about and how best to convey those warnings. Coordinating on-product labels, safety messages in manuals, and safety information in other product-related materials is made much easier if you have a rational, systematic hazard-analysis process in place. The manual, especially, can be a valuable safety resource, providing additional information and explanation for hazards addressed with on-product labels, as well as offering additional safety-related information to the user.

Manufacturers also have a duty to warn users of hazards discovered after the sale. The long life of many products, especially large capital machinery, makes this duty an ongoing concern for manufacturers. For very serious hazards, the manufacturer may choose or be required to perform a recall of the product. For other hazards, manufacturers may choose instead to retrofit additional safety features or offer additional safety-related materials, such as warning labels or safety bulletins. Of course, for either approach to be successful, the manufacturer must be able to find the people who own the products. For custom machinery, doing so may be fairly simple; for consumer products, it is a much bigger challenge.

Whether you are conducting a hazard analysis on a new product under development or designing a new warning for a 30-year-old product, the resulting warning must meet legal and industry standards and communicate to the user. The next chapter looks in more depth at creating effective warnings.

CHECKLIST: MANAGING PRODUCT SAFETY

- Have we conducted a hazard analysis?
- Have we used the hazard analysis to identify needed warnings?
- Have we determined which warnings should be on-product labels and which should be covered in the manual?
- Do we have a system in place to ensure that on-product labels and in-manual labels are consistent?
- Have we identified any post-sale warning issues?
- If so, have we developed a plan to address our post-sale duty to warn?

NOTES

1. *Restatement of the Law Third, Torts: Products Liability* (American Law Institute Publishers, 1998).
2. Law No. 85 (1994). See also an excellent article in *Lexology* at www.lexology.com/library/detail.aspx?g=e013ff26-a955-4faa-bba0-5b7be84baf4f, accessed March 25, 2019.
3. See, for example, *Litigation Forecast 2019* (Washington, D.C. [and other locations]: Crowell & Moring LLP, 2019), www.crowell.com/files/Litigation-Forecast-2019-Crowell-Moring.pdf, accessed March 26, 2019.
4. *Facts and Statistics: Litigiousness*, Insurance Information Institute, www.iii.org/media/facts/statsbyissue/litigiousness/.
5. See "In Tests of Boeing Jet, Pilots Had 40 Seconds to Fix Error," *The New York Times*, March 25, 2019.
6. *Restatement Third*, §2, Comment i.
7. Results from Martindale-Nolo Research's 2017 personal injury study, www.lawyers.com/legal-info/personal-injury/personal-injury-basics/personal-injury-how-much-can-i-expect-to-get.html, accessed March 28, 2019.
8. *Russell v. Mississippi Valley Gas*, Miss. Hinds County Circuit Court, No. 31,510, March 20, 1987.
9. *Restatement Third*, §2, Comment m.
10. *Hubbard-Hall Chemical Company v. Silverman*, 340 F 2d 402 (1st Cir. 1965).
11. *Federal Insecticide, Fungicide, and Rodenticide Act, 40 Code of Federal Regulations* 156.64 (a) (1).

12. The most common situation in which hazards become apparent after the sale is with pharmaceuticals, where side effects may not be apparent until many years later. For example, the drug thalidomide, used in the past to treat morning sickness during pregnancy, was later found to cause birth defects.

13. *Restatement Third*, §2 j.

14. Jessie Y. C. Chen, Richard D. Gilson, and Mustafa Mouloua, "Perceived Risk Dilution with Multiple Warnings," in *Human Factors Perspectives on Warnings, Volume 2: Selections from Human Factors and Ergonomics Society Annual Meeting 1994–2000,* ed. M. S. Wogalter, S. L. Young, and K. R. Laughery Sr. (Santa Monica, CA: Human Factors and Ergonomics Society, 2001), pp. 50–54.

15. When used as a signal word, *NOTICE* is italicized. The other signal words are not.

16. ANSI Z535.4–2011(R2017), § 4.14.

17. *ANSI Z535.6 for Product Safety Information in Product Manuals, Instructions, and Other Collateral Materials* (Rosslyn, VA: National Electrical Manufacturers Association, 2011).

18. *Restatement Third*, §10 a.

19. *Restatement Third*, §10 b.

20. *Restatement Third*, §10 a.

9 Warnings That Work

OVERVIEW

If you can design a hazard out of a product, you have no need to warn, because nothing is left to warn about. But not all hazards can be designed out. And even if a hazard can be guarded, you may still need to warn against removing the guard or operating the machine without the guard in place. Most products have at least some residual hazards that require warnings. How can you be sure your product's warnings work effectively to help keep your users safe? This chapter will help you answer that question. While most of the information in this chapter is presented in terms of on-product warnings (usually labels), much of the information is applicable as well to warnings used in other contexts, including safety bulletins, manuals, packaging, and other product materials.

We will look at what makes a warning legally adequate, what goes into a properly designed warning label, how you can be sure that your warning labels meet applicable standards, and finally, how to test your warnings to be sure they are effective. The first step is to understand the legal purpose of a warning.

WHAT SHOULD A WARNING DO?

While the answer to the question of what a warning should do may seem obvious at first glance (duh, warn people!), there really are differences of opinion about the purpose of a warning. If you ask a human factors expert about the function of a warning label, you will generally get an answer that focuses on whether the warning changes the user's behavior. For example, consider this excerpt from the opening paper in a collection of human factors articles on warnings: "By an effective warning, we mean one that changes behavior in a way that results in a net reduction in the relevant negative consequences."[1] The difficulty with this definition from the manufacturer's perspective is that you don't control the user's behavior. You can design a state-of-the art warning, but the user may deliberately choose to ignore it and do the unsafe action anyway. You have probably done this yourself at one time or another—chosen to turn off, but not to unplug the vacuum cleaner before pulling a partly eaten throw rug out of the brushes, used an adaptor rather than finding a grounded receptacle to plug in a three-prong electrical plug, or gone without hearing protection when operating a circular saw. You may even have said to yourself, "I know this is probably stupid, but it'll only be a minute, and I'll be careful." Does that mean the warning was ineffective? A human factors expert would say so. So, when you suffer an electrical shock because you used an adaptor, is the manufacturer liable for a defective product?

WHAT MAKES A WARNING ADEQUATE?

Fortunately for manufacturers, the legal test for an effective warning is not the same as the human factors test. The legal requirement is this: " . . . sellers must provide

reasonable instructions and warnings about risks of injury posed by products. . . .
Warnings alert users and consumers to the existence and nature of product risks so
that they can prevent harm either by appropriate conduct during use or consumption
or by choosing not to use or consume."[2] In other words, the manufacturer's duty is to
provide information to users, not to change their behavior. Think of it as analogous
to informed consent: when a doctor recommends a medical procedure, he or she is
required to tell you about the potential risks, so you can do a personal cost-benefit
analysis before agreeing to go ahead. The manufacturer has to warn potential users
of the risks involved in using the product.

What does that mean in practical terms? It means that you have to give the user
enough information to make a reasoned decision. While it's impossible to guarantee that a specific warning will be found adequate in court (juries are unpredict-
able), certain elements have emerged as necessary for a warning to be considered
adequate. At a minimum, a hazard warning must do four things:

- Identify the nature and severity of the hazard.
- Tell the user how to avoid the hazard.
- Explain the consequences of not avoiding the hazard.
- Clearly communicate this information to the user.

Let's look at each of these in more detail.

Identify the Nature and Severity of the Hazard

Here's a typical warning statement:

CAUTION: For external use only. May be harmful if swallowed.

This statement tells neither what will happen (the nature of the hazard) nor how
serious the harm will be (severity of the hazard) if the substance is swallowed. It
would be equally applicable to swallowing hydraulic oil or drain cleaner. But the
results would be very different. If you drank hydraulic oil (from a gravel separator,
no doubt), you could suffer "discomfort."[3] In other words, you'd have at most probably a stomach ache and a sore throat. If you swallowed drain cleaner, on the other
hand, you would almost certainly suffer severe and potentially life-threatening corrosive burns to the mouth, esophagus, and stomach.[4]

In some situations, the nature and severity are obvious and need not be explicitly
stated. For example, a warning on a ladder not to overreach need not state, "you could
lose your balance and fall, causing death or severe injury." It's common knowledge
that falling from a height can cause broken bones or worse. Be sure, however, that
what seems like "common knowledge" to you is indeed widely known. To continue
with the ladder example, if your company manufactures ladders, you probably know
that the correct angle for a ladder is about 75 degrees from the ground. This angle
is easily approximated either by making sure that the distance between the base of
the ladder and the support is about one-fourth of the working height, or by standing
upright on the bottom rung and extending your arms. If you can easily grasp a rung

when your arms are straight, that's about the right angle. But people who don't use ladders regularly may not know this, and may set the ladder at too steep or too shallow an angle.

Why don't products have more specific warnings? Usually a vague warning is a result of one of three factors:

- The writer doesn't know any different, and simply follows what he or she has seen on other products.
- The marketing department objects to a specific warning, fearing that it will make the product appear dangerous.
- The legal department objects to a specific warning fearing that if a resulting injury isn't exactly the one warned against, the company will be liable.

The first is easily solved—by the end of this chapter, you'll know exactly what to do.

The second is a bit more difficult. Product safety issues almost always result in competing agendas from different parts of the company. The marketing department is tasked with selling products, and anything that appears to make that more difficult will usually be resisted. If your company establishes a product safety team with authority to make decisions (more on this in Chapter 11), the team will enforce proper warning messages. Even more effective than that, however, you may be able to convince the marketing department that more specific warnings are preferred by consumers. Research indicates that this is in fact the case—because consumers perceive a more explicit warning as suggesting that the company is concerned about the safety of its users.[5]

Objections by the legal department to explicit warnings usually come from a lack of products liability experience—the "rules of the game" for creating legally defensible warnings are very different from those for creating many other kinds of legal documents, and unless your company's attorneys have products liability experience, they may not realize the differences. You may want to recommend that the company consult with an attorney who specializes in products liability to review warnings or to provide training to in-house counsel and technical writing staff. In any case, remember that the legal standard for warnings, as it is for many other aspects of product safety, is *reasonableness*. The manufacturer is required to provide reasonable warnings—not all-inclusive warnings that identify every imaginable kind of harm that could result from a given hazard.

How can you develop a specific warning message? Ask yourself two questions, "What will happen if the user encounters the hazard?" and "How bad will it be?" Use the answers to guide your message. Take, for example, a warning that reads, "Breathing vapors may be harmful to your health." The answer to "What will happen?" might in one case be, "You'll feel light-headed," and in another, "You'll have respiratory problems." If the result of breathing the vapors will be respiratory problems, those could range from mild irritation of the nasal passages and throat for one product to sudden respiratory arrest and death for another. Make sure you tell your users both the nature of the hazard (what could happen) and the severity (how bad it could be).

Tell How to Avoid the Hazard

Oftentimes, warnings will omit this part or make it so vague as to be meaningless. How often, for example, have you seen the advice on a warning label to "Use with adequate ventilation?" That common phrase conveys no real information. What, exactly, is adequate ventilation? Does it mean that I should have a window or two open? Should I have a fan? Can I use this only outdoors? Do I need a fume hood? Do I need to wear a positive-pressure, supplied-air respirator? Depending on the product, any one of those might be the appropriate avoidance action. As you write warnings for your product, try to give the user clear, specific information for how best to avoid the hazard. Instead of saying, "Avoid breathing dust," for example, you could tell the user what kind of respirator to wear or to wet down dust before cleaning it up, as appropriate for the product.

Explain the Consequences of Not Avoiding the Hazard

What's the difference between explaining the consequences of interacting with the hazard and identifying the nature and severity of the hazard? In some cases, they may be one and the same. For example, if the nature and severity of the hazard is identified as electrocution—well, that's the consequence as well. In other cases, it's not so clear. For example, certain machinery may pose an entanglement hazard—but that could simply mean that clothing or loose hair could get caught in a mechanism, resulting in the machine jamming and perhaps damaging clothing or pulling some hair—or it could mean that your entire arm could be drawn into machinery and amputated. Whenever the consequence isn't immediately obvious from the description of the nature and severity of the hazard, be sure to explain the potential consequence of coming into contact with the hazard—in specific terms. Avoid using a stock phrase at the end of each warning: "Failure to follow this warning may result in injury." That kind of generic phrase takes up space and adds no meaning. If there weren't a possibility of injury, why would there be a warning label in the first place?

Clearly Communicate to the User

The previously discussed elements of an adequate warning are more or less straightforward: figuring out in specific terms what the hazard is and telling users how to avoid it. Clearly communicating, however, is a little more complicated. Whether communication is clear or not depends on many factors, including user characteristics. If the user is illiterate, attempting to warn with words alone is certainly not going to communicate clearly. If the user reads only Spanish, presenting an English-language warning will not communicate clearly. Even if your user speaks and reads English, too high a language level may not be clear. For example, "Inhalation of volatile vapors may cause respiratory dysfunction" may be perfectly clear to a medical professional, but not to an ordinary person without a medical background. A better choice might be, "Inhaling vapors may cause difficulty breathing." The next section, "The Anatomy of a Warning Label," offers advice to make the word message more likely to communicate clearly. One of the best ways to ensure clear communication is to focus on keeping your user safe—that will automatically help you communicate clearly.

Focus on Safety Instead of Liability

Many times, manufacturers are understandably very focused on protecting themselves from liability. From a business perspective, this seems on the surface to make sense—million-dollar jury awards can wipe out companies that have no insurance, and drive up the premium cost for those that do. Yet focusing instead on protecting the customer from injury—keeping the user safe—may be a more effective approach. What's the difference? They both have to do with product safety issues, but the goals are subtly different, and sometimes subtle distinctions can have significant consequences. A former colleague is fond of telling the following story: During the Dark Ages when the practice of medicine in Europe was a bit primitive, it could be difficult to tell the difference between deep coma and death. As a consequence, the populace had a considerable (and rational) fear of accidentally being buried alive. This problem was approached differently in England and in Spain. In England, coffin-makers developed elaborate signaling systems, so that if an unfortunate individual happened to wake up while being borne off to the cemetery, he could pull a string placed in his hand inside the coffin and ring a bell on the outside, alerting the pallbearers to the error. In Spain, they simply drove a stake through the heart of the presumed corpse before placing it in the coffin. The two practices had subtly different purposes: in England, they made sure no one who was still alive was buried. In Spain, they made sure that everyone who was buried was dead!

Focusing on liability prevention versus focusing on user safety is a similarly subtle distinction with major effects. If your goal is to write warnings to protect the company from liability claims, your warnings will tend to read like the fine print in insurance policies. Example 9.1 and Example 9.2 show warnings designed to protect the company. These use complex language and try to shift the burden from the company to the user. Contrast them with Example 9.3, which is designed to protect the user. Here the wording is straightforward, and helpful information is included.

Example 9.1: Warning Designed to Protect the Company

Provision of point of operation safety devices consistent with the use and operation of this machine is the sole responsibility of the user of this machine.

Example 9.2: Warning Designed to Protect the Company

All persons authorized to use the equipment must be cognizant of the danger of excessive exposure to x-radiation, and the equipment is sold with the understanding that the ABC company, its agents, and representatives have no responsibility for injury or damage which may result from exposure to x-radiation.

Example 9.3: Warning Designed to Protect the User

WARNING: This product contains chlorine. Mixing this product with other household cleansers, such as toilet bowl cleaners, rust removers, and products containing ammonia, may release deadly chlorine gas. Do not use this product with other chemicals.

Always focus on protecting the user. You will write more effective warnings and put the company in a much more defensible position in the event there is a lawsuit. Remember that one of the elements essential for a successful products liability lawsuit is that there must be injury or damage. If you successfully protect the user, there will be no injury and no liability to worry about. Even if someone is injured and sues, what kind of warning will look better to a jury? One that tries to disclaim responsibility, or one that is clearly aimed at giving users the information needed to stay safe? The fact is, in current law, the manufacturer or seller is liable for defective products, so the disclaimer-style "warning" won't help.[6]

The importance of focusing on protecting the user applies not just to warning labels, but extends to other safety materials as well. Compare Example 9.4 and Example 9.5, both examples of safety information that might be included with instructions accompanying a paint-removal tool, like a scraper or heat gun. Clearly, Example 9.5 is more user-focused. It identifies the hazard and suggests means for minimizing risk. It also points the user to an easily accessible source for more information (a Web page). Example 9.4 buries the identification of the hazard in the middle of a long paragraph and gives only very limited risk-reduction information. The bulk of the text is devoted to explaining why the ABC Company should have no liability if someone suffers lead poisoning after using its tools. Which one would more useful to the user? Which one would you rather defend from the witness stand?

Example 9.4: Safety Instructions Written as a Disclaimer

Important Safety Information

When removing lead-based paint with a heat gun or mechanical removal tools, including pressure washers and/or sand-blasting equipment, a suitable means for collection and disposal of resulting dust, mist, or sludge should be provided, as lead can be absorbed through the skin, ingested, or inhaled as part of dust or mist. Using a heat gun to remove lead-based paint may produce fumes containing lead, inhalation of which can result in exposure to lead. Overexposure to dust, mist, or fumes containing lead can be hazardous to health, particularly if exposure continues over an extended period of time, and may cause temporary or permanent lead toxicity, including irreversible neurological effects. Use of chemical strippers to remove lead-based paint can result in lead-contaminated sludge. Adequate ventilation, appropriate respiratory protection, and eye protection should be provided, and workers should avoid inhalation of and prolonged skin contact with dust, mist, sludge, or fumes. General Industry Safety and Health Regulations, Part 1910, U.S. Department of Labor, published in Title 29 of the *Code of Federal Regulations* and OSHA Lead in Construction Standard Title 29 of the *Code of Federal Regulations*, Part 1926.62, and EPA Lead; Renovation, Repair, and Painting Program, Title 40 of the *Code of Federal Regulations*, Part 745 should be consulted.

Removal tools and methods are only one factor in the hazards posed by lead-based paints. Many variables exist in paint-removal operations, including age and friability of the paint; lead content; means of removal; length of exposure; personal protective gear worn; age and health of the workers; and environmental conditions, including wind direction and speed, temperature, and humidity. A safe paint-removal operation must take all of these variables, and others, into consideration.

ABC Company has no control over the end use of its products or the environment into which those products are placed. ABC urges that its customers adhere to the recommended standards of use of their paint-removal tools, and that they follow procedures that ensure safe paint-removal operations. The technical information included in these instructions as well as recommendations on removal practices referred to herein are only advisory in nature and do not constitute representations or warranties and are not necessarily appropriate for any particular work environment or application.

Example 9.5: Safety Instructions Written to Protect the User

WARNING!

Lead Paint Hazard

Removing lead-based paint produces hazardous material, regardless of the method used to remove the paint. Lead can be absorbed through the skin, swallowed if it comes in contact with food or drink, and inhaled as tiny particles in the air or as fumes produced when paint is heated.

Exposure to lead can cause irreversible brain and nervous system damage, especially in children.

When removing lead-based paint, always:

1. Contain the area where the paint removal is taking place. Keep others out.
2. Wear appropriate personal protective equipment, including skin, eye, and breathing protection.
3. Safely collect, contain, and dispose of dust, mist, or sludge produced.
4. Avoid contact with lead-contaminated material.

Many factors affect the hazards connected with removal of lead-based paints. ABC urges you to learn more about safe work practices or to consult with a certified lead-remediation specialist.

For further information, including safe work practices, visit www.epa.gov/lead.

Why are some warnings written to sound like disclaimers? It may be because the wrong person is writing the warning.

WHO SHOULD WRITE THE WARNINGS?

Because providing adequate warnings is part of a manufacturer's legal duty, in many companies, the legal department insists on writing the warnings. This practice is usually not a good idea, even if your corporate attorneys are very capable. Attorneys, like other professionals, tend to specialize. They become very skilled in the areas of law that they practice regularly, but usually are not top-notch in other areas. You wouldn't ask a cardiologist to do a hip replacement would you? Well, don't expect your corporate counsel to be on the cutting edge of products liability law. As noted, products liability law is complex and rapidly evolving—it's hard to keep up even if you work in that area full time. Moreover, company lawyers spend most of their time drafting language for sales contracts, licensing agreements, and warranties, among other duties—and that may actually make it harder for them to write good warnings.

Think about it—what does a good sales contract or licensing agreement or warranty do? It spells out in excruciating detail how every single imaginable situation will be handled. Even if it is not written in "legalese" (the party of the first part, etc.), it still is likely to have phrases like "including but not limited to," "no part of this agreement shall be construed . . ." and "the company accepts no liability for . . ." While phrases like these may represent effective wording for some kinds of legal documents, they make for terrible warnings. Warnings should be concise, straightforward, easy to understand, and focused on keeping the user safe.

In other companies, responsibility for producing all product documentation, including manuals, spec sheets, sales brochures, etc., is given to the marketing department. While it makes a lot of sense to put all those functions under a single authority, giving marketing the final authority has drawbacks as well. One of these is that, as noted earlier in the chapter, marketing professionals are generally going to want to minimize the number and intensity of warnings—both on the product and in the manuals. Marketing personnel should certainly be involved in product decisions, but they shouldn't be given sole responsibility for writing the warnings—nor the power to tone down the language. For warnings to do their job, the language needs to convey the seriousness and immediacy of the hazard being warned against.

Nor should marketing have veto power over the appearance of on-product warnings. On-product warnings can take a variety of forms, from the familiar stick-on warning labels using color, signal word, and pictorial to help convey the message to words stamped into or embossed on the product housing. Some products use hang tags, box inserts, or packaging to convey warning messages. The decision as to what form warnings should take must consider not only appearance, but also location with respect to the hazard, whether there are mandatory regulations regarding wording or appearance, and the need to conform to industry standards. Marketing should certainly have a voice—but it should not be the only one.

The best approach is to use a team of people representing different areas of the company to conduct the hazard analysis and decide what warnings are needed. How to put together a product safety team is discussed in Chapter 11. Whatever approach you use, you may find it beneficial to contract with an outside warnings consultant to help you develop product warnings. A good consultant will bring a broad perspective to the table. A consultant will know what other companies and industries are doing in this area and can conduct a literature search on safety issues connected with your product. He or she will be aware of current trends in products liability law. And most important, a consultant will bring an outside eye. Just as in the development of instructions, one of the primary obstacles for in-house personnel in addressing product safety is overfamiliarity with the product. Knowing your product inside and out can make it difficult to look at the product as a first-time user might—which is critical to developing good warnings.

MEETING STANDARDS . . . WHICH ONES?

One of the first questions your insurance company will ask when you try to get products liability insurance is, "Does your product meet industry standards?" Certainly, one aspect of that question is whether the warnings on the product meet standards. "Standards" is often used loosely to mean "requirements," which could be government regulations, industry standards, or requirements set by independent testing organizations such as Underwriters Laboratories (UL). What are the requirements for warnings? How do you find out? For established product types, a good source of information is usually your trade association. Whether you manufacture paint or mufflers or woodworking machinery, most likely you can find an association of manufacturers of similar products. Association websites often have links to government regulations as well as standards—in fact, in many cases it is the association that develops the standards.

If you have a new product type or a unique product, finding what requirements apply can be much more challenging. In many cases, there may simply be no regulations or standards specific to your product type. But you may have to spend a good deal of time researching the question before you feel comfortable that none exist. One approach is to research related products to see what kinds of warnings are required and proceed in a similar vein. In any case, the first step is to identify any government regulations that cover your product.

GOVERNMENT REGULATIONS

Government regulations for warnings have the force of law. If your product does not comply with labeling and warning requirements, it cannot legally be sold and the manufacturer and sellers can be fined. There are literally dozens, if not hundreds, of government regulations of various kinds of products. It would be impossible to detail each of them in a book like this. However, three that have broad application to many products are

- The *Federal Hazardous Substances Act*[7]
- The OSHA Hazard Communication Standard[8]
- The *Consumer Product Safety Act*[9]

This section provides a brief introduction to these.

The Federal Hazardous Substances Act

The *Federal Hazardous Substances Act* (FHSA) covers any household product that is

- Toxic
- Corrosive
- An irritant
- A strong sensitizer
- Flammable or combustible
- Generates pressure through decomposition, heat, or other means

and that could cause "substantial injury" in the course of normal or foreseeable handling, *including ingestion by children* [italics added].

As you can imagine, the FHSA covers a lot of products. The act also covers toys that present an electrical, mechanical, or thermal hazard. It does exclude products covered by the *Federal Insecticide, Fungicide, and Rodenticide Act* (FIFRA), as well as food and cosmetics, and various other categories of products, but even so, a quick look through your cleaning and workshop supplies will probably reveal a good many products covered by the FHSA.

What does the FHSA require of the manufacturer or seller in terms of warnings?[10] According to the FHSA, the product label for any hazardous substance as defined under the act must "conspicuously" state the following:

- The name and location of the manufacturer or other seller
- The common name of the hazardous substance(s) in the product
- The appropriate signal word ("DANGER," "WARNING," "CAUTION") for the hazard presented
- A statement of the "principal hazard" (such as "Flammable," "Causes Burns," etc.)
- Avoidance information
- First-aid treatment
- Direction to "Keep out of reach of children"

Additional requirements pertain to "highly toxic" substances and items requiring special handling.

The hazard information required is similar to that required for any good warning: information as to the nature and severity of the hazard and how to avoid it.

The FHSA requires that the necessary labeling be "prominent" on the label. The specifics of how that is to occur are spelled out exhaustively in the *Code of Federal Regulations*.[11] The rules address layout, relationship to other text on the product label, how to manage different sizes and shapes of packaging, and much more. The rules do not require any one specific format, but rather specify criteria that the format must meet. One aspect to bear in mind in ensuring that your warning is conspicuous is where it appears in relation to other text on the label. Hazard information buried in the middle of a long paragraph of general instructions is less likely to be seen than if it were set apart, regardless of whether it is presented in bold type or given some other formatting emphasis.

The OSHA Hazard Communication Standard

"The *OSHA Hazard Communication Standard* is intended to . . . ensure that the hazards of all chemicals produced or imported are evaluated, and that information concerning their hazards is transmitted to employers and employees. This transmittal of information is to be accomplished by means of comprehensive hazard communication programs, which are to include container labeling and other forms of warning, Material Safety Data Sheets, and employee training."[12] All OSHA regulations apply to employers and must be followed in the workplace, but they do not cover non-work-related environments. In other words, the labeling requirements in the *OSHA Hazard Communication Standard* do not address consumer product labeling at all. However, if you produce chemicals or incorporate them in your products, you must follow the standard in order to get your products to distributors and other sellers.

For example, suppose you manufacture drill bit blanks that you then sell to a drill-bit manufacturer. The blanks are ground into various sizes and styles of bits by that drill-bit manufacturer and then resold to the end-user. The blanks are made from metal powders containing tungsten carbide, with cobalt as a binder. Inhaling dust containing these elements can cause "hard-metal lung disease"—an irreversible condition that can lead to lung fibrosis and death.[13] What are your obligations as the manufacturer of the blanks to warn under OSHA's Hazard Communication Standard?

First, you must provide a label for each container that:

- Identifies the chemical
- Provides an "appropriate hazard warning"
- Provides the name and address of the manufacturer (or other responsible party)

Second, you must provide a Safety Data Sheet (SDS) for each chemical. (Note that these used to be called Material Safety Data Sheets, or MSDS.) The SDS must contain the following information:[14,15]

1. *Identification* of the product, including chemical, manufacturer contact information and recommended uses
2. *Hazard(s) Identification* including hazard classification (e.g., flammable liquid), and avoidance information
3. *Composition/Information on Ingredients*, including chemical name(s) and synonyms and concentration
4. *First-Aid Measures*
5. *Fire-Fighting Measures*, including recommended extinguishing agents and other specific fire-related hazards
6. *Accidental Release Measures*
7. *Handling and Storage*
8. *Exposure Controls/Personal Protection*, including OSHA Permissible Exposure Limits (PELs) and other measures
9. *Physical and Chemical Properties*
10. *Stability and Reactivity*
11. *Toxicological Information*
12. *Ecological Information (non-mandatory)*
13. *Disposal Considerations (non-mandatory)*
14. *Transport Information (non-mandatory)*
15. *Regulatory Information (non-mandatory)*
16. *Other information*, such as date of last revision or other useful information

Sounds pretty comprehensive, doesn't it? Unfortunately, the reality does not always meet the promise. Safety Data Sheets vary enormously in the quantity and quality of information provided. Example 9.6 shows the health hazard information provided in the SDS for tungsten from four different manufacturers. As you can see, not only is the level of detail different, but the assessment of the degree of hazard presented by tungsten is also different—which would certainly affect the information you would provide on the package label.

Example 9.6: Sample Hazard Information from Four Safety Data Sheets for Tungsten

Manufacturer A

Effect of Overexposure: dust, mist, and fumes generated during physical or metallurgical treatment may cause mild irritation of the nose and throat. With the exception of two Russian studies that found early signs of pulmonary fibrosis in some workers exposed to tungsten trioxide, tungsten metal, and tungsten carbide, most studies have shown tungsten to be toxicologically inert. Skin and eye contact may cause irritation due to abrasive action of the dust. Current scientific evidence indicates no adverse effects are likely from accidental ingestion of small amounts of tungsten.

Manufacturer B

- EYE CONTACT: The dust causes slight to mild irritation of the eyes.
- SKIN CONTACT: The dust causes slight to mild irritation of the skin.
- INHALATION: Irritating to the respiratory tract.
- INGESTION (Acute): None
- INGESTION (Chronic): Large overdoses may cause nervous system disturbances, and diarrhea.

Manufacturer C

- This chemical is considered hazardous by the 2012 OSHA Hazard Communication Standard (29 CFR 1910.1200)
- Label Elements Signal Word: Warning
- Hazard Statements:

  - Flammable solid
  - May form combustible dust concentrations in air
  - Flammable solids Category 2
  - Combustible dust Yes

Manufacturer D

- HAZARD STATEMENT

  - Causes respiratory tract, eye, and skin irritation.
- PRECAUTIONARY MEASURES

  - Avoid contact with eyes, skin, and clothing. Do not eat, drink or smoke when using this product. Keep container closed, use protective equipment as required and wash thoroughly after handling.

- POTENTIAL HEALTH EFFECTS
  - EYES: Irritation possible
  - SKIN: Irritation possible
  - INHALATION: None
  - INGESTION: Gastrointestinal irritation possible
  - MEDICAL CONDITIONS AGGRAVATED: Exposure to tungsten dust or fume may aggravate chronic respiratory disease
  - CHRONIC OVEREXPOSURE: Not known
  - ACUTE: OVEREXPOSURE: Not known
  - CARCINOGENICITY: None

Providing an SDS (normally, you would use the one provided by your supplier) and a package label fulfills your legal obligation to warn under the terms of OSHA's Hazard Communication Standard—but does it completely fulfill your duty to warn? An SDS and a label would certainly identify the nature and severity of the hazard, explain how to avoid the hazard, and what the consequences of interacting with the hazard would be (accurately or not!), but what about the fourth element of an effective warning? Would it clearly communicate this information to the user? As usual, that depends on various factors, including user characteristics. The standard requires that manufacturers provide employers with an SDS, and that employers make the SDS available to employees. But what if the employees can't read English? Or they can't read English at more than a fifth-grade level? Would an SDS mean much to them?

One of the challenges for manufacturers is to find ways to reach all the users of their products with safety information. Doing so is difficult enough when the buyer is the end-user of the product—it's even more difficult when the buyer is another manufacturer using the product as a component or raw material for their products.

Consumer Product Safety Act

The *Consumer Product Safety Act* (CPSA) establishes the Consumer Product Safety Commission (CPSC) and gives it authority to develop rules to implement the act. Under the CPSA, the Commission can ban hazardous products, set safety standards for products, and require reporting and inspections relating to product safety. The CPSC has established safety standards for many products, from architectural glazing material to baby pacifiers. Some of these standards include labeling requirements. For example, the Safety Standard for Walk-Behind Power Lawn Mowers[16] prescribes the specific label, shown in Figure 9.1, to be placed near the blade housing. Other CPSC safety standards prescribe certain language, but do not specify pictorials or placement—see for example, the standard for residential garage doors.[17]

The CPSC does not promulgate safety standards for all products. In fact, under the act,

"The Commission shall rely upon voluntary consumer product safety standards rather than promulgate a consumer product safety standard prescribing requirements described in subsection (a) whenever compliance with such voluntary standards would eliminate or adequately reduce the risk of injury addressed and it is likely that there will be substantial compliance with such voluntary standards."[18]

FIGURE 9.1   Required warning on lawn mowers.

Source: From *Code of Federal Regulations*, 1205.6 (a) Safety Standard for Walk-Behind Power Lawn Mowers, Consumer Product Safety Commission.

In many cases, the CPSC actively works with and supports voluntary standards-setting activities.

VOLUNTARY STANDARDS

Manufacturers' associations and independent standards organizations are the primary groups that develop safety standards for products. There are literally hundreds of voluntary product standards covering all aspects of design, manufacturing, and labeling for all kinds of products. It is obviously impossible to list them all here, but you should become familiar with standards that apply to your industry. Although these standards are technically voluntary, often they become a *de facto* benchmark for product safety. Additionally, adherence to these standards may be required by codes or other regulations that have the force of law, essentially turning the standards into indirect legal requirements.

Another type of standard is created by independent testing and certification organizations. Probably the best known of these organizations is Underwriters Laboratories, Inc. (UL). In addition to conducting testing of products—and providing certification to those that pass muster—UL also develops safety standards of its own. These are recognized—and even may be required—by many regulatory agencies.

If your product is covered by a regulation or certification requirement that specifies certain warning language, of course you must comply. But while regulations may specify warnings that a manufacturer must include, they generally do not *prohibit* additional warnings. If your hazard analysis suggests that additional warnings would be useful, then by all means, include them. What forms should those warnings take? The next section addresses this question.

THE ANATOMY OF A WARNING

Until near the turn of the 21st century, no single format for hazard warnings dominated. OSHA had requirements for workplace safety signs, but these did not apply to on-product warning labels. No standard existed for warning statements in manuals. In 1991, the first edition of ANSI Z535.4 *Standard for Product Safety Signs and Labels* was published. The comparable international standard (ISO 3864) was first published in 2002. Not until 2006 was there a standard (ANSI Z535.6) for warnings in manuals. These standards are regularly revised. In the United States and Canada, the most common format for on-product warning labels is the one specified in ANSI Z535.4. Internationally, the format specified by ISO 3864–2 is more commonly followed. We will discuss both formats, as well as noting that recent revisions make it easier for manufacturers to develop effective formats that meet both standards. We begin with ANSI Z535.4.

ANSI Z535.4

The ANSI Z535.4 standard[19] outlines a format using five elements to convey hazard information as shown in Figure 9.2:

- Safety-alert symbol: indicates a hazard that may result in personal injury
- Signal word: conveys severity of the hazard
- Color: conveys severity of the hazard
- Pictorial (optional): typically conveys either the nature of the hazard or avoidance information
- Word message: conveys nature and severity of the hazard and avoidance information

Safety-Alert Symbol

As noted in Chapter 8, the safety-alert symbol consists of an exclamation point surrounded by an equilateral triangle. The colors used vary depending on the severity of the hazard.

Signal Word

Choosing the correct signal word depends on the results of your hazard analysis. The meanings of the signal words are as follows:

- "DANGER" indicates a hazardous situation, which, if not avoided, will result in death or serious injury.

- "WARNING" indicates a hazardous situation, which, if not avoided, could result in death or serious injury.
- "CAUTION" indicates a hazardous situation which, if not avoided, could result in minor or moderate injury.
- *"NOTICE"* is used to address practices not related to personal injury.

As noted in Chapter 8, the only difference between "DANGER" and "WARNING" is the likelihood of occurrence. Both are used for hazards that have the potential for causing death or severe injury, but if the signal word is "DANGER," then encountering the hazard means that injury (at least) is a near certainty, not just a possibility. "DANGER" and "WARNING" and "CAUTION" are always accompanied by the safety-alert symbol. *"NOTICE"* (without the safety-alert symbol) is used to address practices that do not have a personal injury component—for example, a behavior that could cause damage to the product.

FIGURE 9.2 Sample ANSI-style warning label. This warning label contains the signal word panel with safety-alert symbol and associated color, a hazard pictorial, and a word message.

Source: Copyright © Clarion Safety Systems, Inc. With permission.

Your hazard analysis should be your guide in choosing the correct signal word. If you use a single label to address more than one hazard, always use the signal word that is appropriate for the more serious hazard. Of course, the decision to use one signal word over another may still require a judgment call. For example, where do you draw the line between "moderate" injury and "severe" injury? Most people would probably agree that an amputated arm would be a severe injury, but what about a crushed finger? How likely must an injury be for the hazard warning to rate a "DANGER" signal word? The standard includes a decision chart to help you make those decisions. It's interesting to note that the standard specifies that "DANGER" is to be "limited to the most extreme situations." Doing so helps to prevent overuse of "DANGER" ("just to be sure we're covered") with the consequent erosion of its intensity.

Using your hazard analysis to guide you in selecting the correct signal words will also help ensure that you use them consistently. If a pinch point on the right side of the machine rates a "WARNING" signal word, then the label for a similar pinch point on the left side should also use "WARNING." Be sure that your use of signal words is consistent throughout all the material pertaining to the product. Often the manual is written by one person and the warnings are created by another—but the manual contains safety messages as well. Be sure that the signal words are consistent wherever a given hazard is addressed.

Color

Each signal word is associated with a specific color, as shown in Figure 9.3. Note that these are special safety colors that are specified in ANSI Z535.1 American National Standard for Safety Colors.

Pictorial

The use of pictorials (also called safety symbols) in warning labels has become much more common, although not universal. The advantage, of course, is that a pictorial representation of a hazard or user behavior (such as wearing of safety glasses) can be understood by those who cannot read the word message. Naturally, this means that any pictorials used in warning labels must be understandable without additional explanatory words. Probably in an effort to ensure instant recognition, over the course of the last 40 years or so, the pictorials used in warning labels have become increasingly standardized—although the systems of symbols used vary geographically. Typically, in the United States and Canada, we see a stylized silhouette of a person or body part, as in the label shown in Figure 9.2. This silhouette, sometimes referred to as "flatman," became widely used beginning in the 1970s and 1980s. FMC Corporation developed a kit for creating warning labels that included master art for a number of hazards and instructions for designing new pictorials using the same system,[20] and Westinghouse Electric Corporation also published a handbook for creating labels.[21] See Figure 9.4 and Figure 9.5 for samples of the label artwork from those two books.

FIGURE 9.3   Signal words and color panel from ANSI Z535.4. Each signal word has a corresponding color to indicate severity and likelihood of harm.

Source: Reprinted by permission of the National Electrical Manufacturers Association from ANSI Z535.4–2011 (R2017).

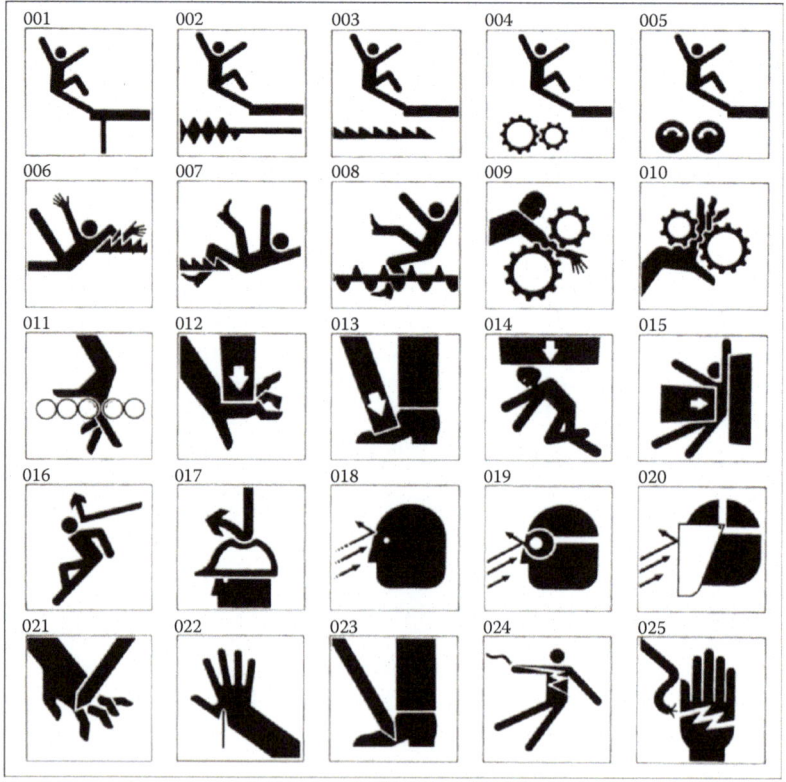

FIGURE 9.4   Sample pictorials.

Source: From *Product Safety Sign and Label System*, 3rd ed., FMC Corp., Santa Clara, CA, 1980, p. 7–2. With permission.

⚠**DANGER**

Hazardous voltage.
Will cause
severe injury or death.

Do not open this cover
if blower is running.

See instruction book .

NOTICE

Connect
thermostat leads
in series
with Stop Button
of 3-wire pilot circuit
in motor controller.

⚠**WARNING**

Hazardous voltage.
Can shock, burn,
or cause death.

Turn off power
before inserting
maintenance handle.

⚠**WARNING**

Hazardous gas.
Can sting eyes,
irritate nose,
or cause death.

Ventilate breaker
before entering.
See instruction book .

FIGURE 9.5   Sample warning labels.

Source: From *Product Safety Label Handbook*, Westinghouse Electric Corp., Pittsburgh, PA, 1981.
With permission.

Flatman-style pictorials are not the only graphical symbols used in warning labels. The United Nations, for example, has published a document including a system of symbols for chemicals.[22]

Whether you use an already developed symbol or create your own pictorial, you must ensure that is effective at conveying the hazard. Two critical guidelines to follow are these:

- Keep the pictorial simple.
- Make the meaning intuitive or train your users.

A pictorial for a warning label must be extremely simple because the goal is to communicate the hazard (or avoidance information) at a glance, so to speak. For a warning to be effective in conveying the information that the user needs to avoid a hazard, it must do so rapidly—before it's too late and the user has been injured. A complex drawing takes too long to interpret. Keep the elements of the graphic to the absolute minimum needed. Once again, as in other aspects of manual development, form follows function.

Any graphical image is an abstraction, and may or may not be immediately understandable. An intuitive symbol, like a human figure or human hand, is recognizable without training. After all, people have been drawing hands since prehistoric times. By contrast, a purely abstract symbol, such as the biohazard symbol (Figure 9.6), requires

FIGURE 9.6   Biohazard label. This is an example of an invented symbol that is not intended to be intuitively understood, but rather learned through training. For that reason, the goal was to avoid using a symbol that had other associations.

Source: From *29 Code of Federal Regulations*, §1910.1030 (g)(1)(i)(B) Bloodborne Pathogens. Occupational Safety and Health Administration.

that users be trained to understand it. Without training, nothing intuitively connects the symbol to the meaning.[23] Suppose your pictorial is somewhere in between? How can you be sure it's comprehensible? The best way is to test it on people similar to the expected users. The last section of this chapter explains how to test warnings.

Word Message

Writing the word message for a label is an exercise in concision—the art of getting your meaning across in as few words as possible. The trick is that getting your meaning across is just as important as using as few words as possible. Rarely, however, are word messages on warning labels too short—although they may be too vague. Here are some guidelines for writing—and designing—effective word messages:

- Use simple, easy-to-understand language.
- Be specific.
- Use active voice and command verbs.
- Format text left-justified and use "sentence" case.

Use Simple Language

It's always proper in product safety contexts to use simple, direct language, but it's crucial on warning labels. Just as no one willingly sits and reads an operator manual from cover to cover for the sheer joy of it, no one studies warning labels. People glance at labels—sometimes while they are already moving toward a hazard. The label needs to communicate both what the hazard is and how to avoid it, all in a split second. As a result, when you draft the word message, expect to go through successive drafts as you hone it to its essence. Your process might look like the following series of drafts to develop wording for a label:

1. Exercise caution working around gears. Your hands could become entangled and your fingers could be cut or crushed.
2. Use care around gears. If your hand is entangled, your fingers could be cut or crushed.
3. Keep hands away from gears. They could be entangled and cut or crushed.
4. Entanglement and crush hazard. Keep hands away from gears.

Notice that the fourth iteration reversed the order of the two parts of the message. Which should you put first—the hazard identification or the avoidance information? Identifying the hazard first provides a context for the avoidance information that may make it more compelling. On the other hand, if you have any concern that users might not have time to read through the hazard identification before it's too late and they've already encountered the hazard, put the avoidance information first: Keep out. High voltage.

Always use the simplest language that will accurately convey the message. In a warning about asbestos, instead of using the word "mesothelioma," use "lung cancer." While mesothelioma is more specific and precise to asbestos-caused disease, it's also more obscure and not likely to be familiar to workers who encounter asbestos. In a warning about the hazard of skin injection of fluid under pressure, instead of saying the affected tissue must be "excised," say that it must be "surgically removed." Why not say "cut out?" It would be simpler and readily understood, but it might not make clear that it's not a do-it-yourself procedure.

Be Specific

Be specific not only about the hazard itself, but also about avoidance or preventive information. In particular, avoid weasel words (see Chapter 4) like these:

- Avoid excessive exposure. (How much is that?)
- Check tightness periodically. (How often is that?)
- Wear appropriate personal protective equipment. (Such as?)
- Keep a safe distance away. (How far is that?)
- Lower boom slowly. (How long should it take?)

Provide specific information about what exactly to do to avoid the hazard. Like weasel words, some phrases that seem to convey avoidance information may actually give the user precious little help. For example, a fairly common directive is this: "Use extreme caution when . . ." What exactly does that mean? What should I do when I use extreme caution as compared to when I exercise ordinary caution—or even none at all? Do I move more slowly, or more quickly, whisper, or what?

Use Active Voice and Command Verbs

Just as active voice and command-form verbs (imperatives) are good style for instructions, so they are for warnings—and they save words. It's a lot shorter to say, "Lock boom before transport," than to say, "Boom must be locked before unit is transported." As with instructions, using command verbs also makes it clear who is to do the action. Avoid using the word "recommend" or "recommended" in warnings, as in "WARNING: Checking tension daily is recommended."[24] The word "recommend" suggests that it's a good idea, but it's optional. Any avoidance action listed in a warning should not be optional—it should be mandatory. Put the advice in the form of a clear command: "Check tension daily."

On the other hand, avoid telling people to do something you know they are unlikely to do. For example, if your company manufactures tires, you might reasonably direct consumers to check their car's tire pressures monthly. But it would be unreasonable to direct them to check it every day—no one will do this. Just as a silly warning ("Do not iron clothes while wearing them.") can result in users ignoring more important warnings, unrealistic avoidance information may result in users ignoring other avoidance directives—and it probably will not protect your company from liability.

Format Text as Left-Justified and in Sentence Case

The most readable format for safety warnings (just as for regular text) is left-justified and ragged right. As discussed in Chapter 4, the eye will track better with the ragged right edge. Never center the text line by line—which is sometimes done in a misguided effort to make the warning stand out. Use ordinary "sentence case," that is, initial capital on the first word of the sentence and the rest lowercase. While it is certainly acceptable to put occasional words in upper case to emphasize them, putting the entire word message in uppercase makes it harder to read. It is also acceptable to use bullets for individual points in the word message.

ISO 3864–2

The primary international standard for communicating hazard information is ISO 3864–2 *Graphical symbols—Safety colours and safety signs, Part 2: Design principles for product safety labels*, published by the International Organization for Standardization.[25]

This standard relies much more heavily on visual elements such as color, shape, and symbols than on text to communicate hazard information. In fact, labels without any word message—even a signal word—are acceptable under the standard. The standard uses three different shape/color combinations with associated pictorial symbols to convey different types of information:

- Warning information uses a yellow triangle.
- Prohibited actions use a red circle-with-a-bar.
- Mandatory actions use a blue circle.

Figure 9.7 shows how these might be used to convey information about a crush hazard, including that user should not put his or her hand into machinery and should read the instructions before using the product.

It makes good sense for an international standard to rely heavily on visual elements—they do not need to be translated. On the other hand, they also depend on the user's being able to interpret the symbol. The standard recognizes this and recommends use of the manual to provide information about what each of the label elements means. A related standard, ISO 7010 *Graphical Symbols—Safety Colours and Safety Signs—Registered Safety Signs* provides a variety of "official" symbols for use on warning labels.

FIGURE 9.7   ISO-style warning labels using pictorials only.

HARMONIZATION OF ANSI Z535.4 AND ISO 3864–2

What if you sell both to the United States and overseas? Do you need two sets of labels, one ANSI-style for use in the U.S. and another ISO-style for the rest of the world? ANSI Z535.4 now incorporates ISO 3864–2 by reference, meaning that a label conforming to ISO 3864–2 also conforms to ANSI Z535.4, so you could simply follow the ISO symbol-only standard for all your labels, since to do so would mean that you were also in compliance with ANSI. That is a legitimate option, but it does have a downside. Many of the ISO symbols laid out in ISO 7010 are not yet well known in the United States and might not convey the intended message.

Fortunately, efforts have been ongoing for several years to harmonize the two standards. ANSI Z535.4 already incorporates ISO 3864–2 by reference, but that's only the beginning. ANSI Z535.4 also allows formatting the safety-alert symbol on a yellow background with a black border,[26] and enclosing the pictorial in a similar yellow triangle that equates to the ISO general warning sign.

Changes have also taken place in the ISO standard. The 2016 version of ISO 3864 includes several new provisions:

- Signal words (DANGER, WARNING, CAUTION) may be used on labels. These signal words have the same definitions as in ANSI Z535.4.
- A "hazard severity panel" containing the general warning sign with or without a signal word may be used. This panel is rectangular and is color-coded to reflect severity (red, orange, or yellow).
- Supplemental safety information in text form may be used.

The result is that it is now possible to design a label in familiar ANSI-style format that also conforms to ISO. See Figure 9.8.

Regardless of which format you choose to follow, you should ensure that the label conveys the information necessary to keep the user safe.

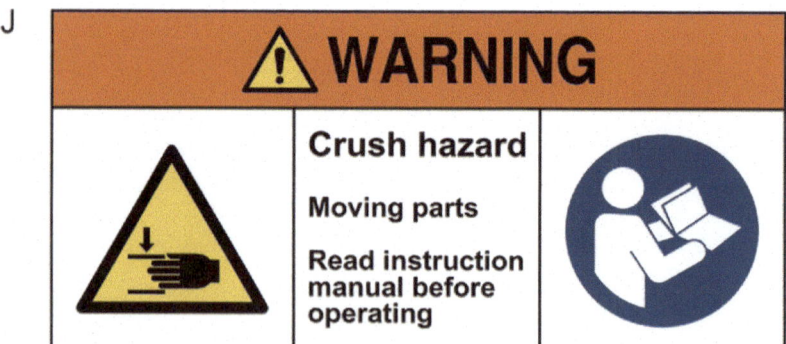

FIGURE 9.8   Harmonized warning label meeting ISO 3864–2 using familiar ANSI-style elements.

Source: ©ISO. This material is reproduced from ISO 3864–2:2016, with permission of the American National Standards Institute (ANSI) on behalf of the International Organization for Standardization. All rights reserved.

ANSI Z535.6

In 2002, the ANSI Z535 Accredited Standards Committee voted to form a new subcommittee to develop a standard for hazard communication in manuals and other product-related materials that would complement the other members of the ANSI Z535 "family" of standards. The result is ANSI Z535.6 *For Product Safety Information in Product Manuals, Instructions, and Other Collateral Materials*. Like the other ANSI Z535 standards, adherence is voluntary. However, just as the ANSI Z535.4 standard for product labels has become so widely accepted that adherence is expected, the same is becoming true for ANSI Z535.6.

This newest addition to the ANSI Z535 family of standards brings some organization to the presentation of safety messages in manuals, but at the same time it allows for a great deal of flexibility. This brief introduction will explain:

- What the standard covers—and what it does not cover
- How the standard organizes safety information
- How the standard can help you with your safety information

For specific guidance about inserting safety information into your own product manuals, you will need to consult the standard directly.

What ANSI Z535.6 Covers

The new standard covers conventional printed documents (or printable electronic documents) providing instructional information about product use, maintenance, and repair. Everything we typically would think of as a "manual" or "instruction sheet," including service and repair manuals, installation instructions, and quick-start guides, is covered. It also includes instructions that appear on packaging. But it does not include everything. The standard (§ 4.2) specifically excludes the following:

- Product signs or labels (these are covered by ANSI Z535.4)
- Material Safety Data Sheets (MSDS)
- Advertising and promotional materials
- Anything not print-based (e.g., videos)

While primarily concerned with documents that are text-based, it may also have application to manuals and similar materials that are mostly pictorial (§ 3.1).

Categories of Safety Information

The standard organizes safety messages into four categories:

- Supplemental directives
- Grouped safety messages
- Section safety messages
- Embedded safety messages

The definitions of these can be a little confusing, but here is a brief explanation with examples.

Supplemental directives are safety-related, but they do not address specific hazards. An example might be a statement at the beginning of the manual urging the reader to "Keep this manual for future reference," or a heading on the "safety page" directing the reader to "Read all warnings before using product." The key feature of supplemental directives is that they do not tell about specific hazards, but rather deal with safety in a general way.

Grouped safety messages are just that—safety messages that appear in a group. The traditional "safety page" at the front of a manual falls under this definition. Also covered under grouped safety messages would be a separate manual dealing exclusively with safety, such as a publication for the printing industry on safe work practices around printing presses.

Section safety messages are those that apply to an entire chapter or subsection of a manual. These may be presented all together at the beginning of the section (but they do not thereby become "grouped safety messages").

Embedded safety messages are those that apply to a *specific* procedure or step in a procedure and that appear (ideally) just before the text to which they apply.

The lines between these can sometimes blur. As with any new attempt to categorize information from a wide variety of contexts, sometimes it can be difficult to figure out whether a safety message should be a section safety message or an embedded safety message. Note also that the categories are not mutually exclusive: just because a given warning appears as a section safety message at the beginning of a section doesn't mean that you cannot also include that same message as an embedded warning as well. If a user might jump directly to a procedure without reading the first part of the section, doing so might be a very good idea.

How ANSI Z535.6 Can Help You

Let's begin with what the standard will not do. It won't tell you when you need a warning, nor will it tell you what that warning should say. No general standard could do that, simply because of the variety of products (and hazards) and the variability in the use environments. While it might be tempting to say that every chain drive needs a warning about entanglement, it's simply not true: if the chain drive is only accessible when much of the machine is dismantled (and therefore not functional), the potential for entanglement is pretty low.

What the standard does do is help you develop a consistent format for different kinds of messages. ANSI Z535.6 uses the same system as ANSI Z535.4 in terms of signal word definitions, associated colors (although color is not required), and the use of the safety-alert symbol. For each type of safety message, the standard prescribes a format or choice of formats. Without going into each example, for which you should consult the standard itself, some general principles emerge:

- Grouped safety messages should be organized.
- Grouped safety messages should not look like a collection of labels.
- Section safety messages should look different from ordinary text.
- Embedded safety messages may or may not look different from ordinary text.

As noted earlier, part of the reason that people often skip over the safety pages at the beginning of a manual is that the messages are not presented in a coherent way and may seem irrelevant. The ANSI Z535.6 standard recommends dividing grouped safety messages into logical categories to increase the likelihood that users will read them.

At the same time, your grouped safety messages should not look like a collection of product labels. Formatting each message with a signal word panel and hazard pictorial and surrounding the entire thing with a box does nothing to encourage reading. While it is acceptable to use a box or similar format device to emphasize a particular warning in a group of warnings, if you use the same device for every message, none stands out.

Section safety messages, which usually appear at the beginning of a chapter or other section, should look different from ordinary text. You can use different techniques to accomplish this. You should use either the safety-alert symbol or a signal word panel to identify the messages as relating to safety (although you do not need to use the device with each message if there are several together). You may use bold type or some other means to distinguish the text of the message from ordinary instructions.

On the other hand, embedded safety messages may appear without special formatting—although they may be accompanied by a signal word, a safety-alert symbol, or use of some other formatting device. The idea with embedded safety messages is that the message will get to the user because the user is reading the procedure. If you use special formatting or put a box around the safety message, it may interrupt the flow of information too much and cause the user to skip over it.

The ANSI Z535.6 standard can help you sort out the warnings that appear in your manual and ensure that they follow a rational and consistent framework to manage your duty to warn.

OTHER CONSIDERATIONS FOR ON-PRODUCT LABELS

DURABILITY

On-product warnings must be designed to last the life of the product. If labels fade to illegibility or fall off entirely, they can no longer be effective at providing the user needed hazard information. Of course, the longer the life of the product, the more difficult it is to design on-product warnings that will last, and the less likely it is that the user will have access to other sources of hazard information, such as the original packaging or the product manual or even an electronic copy for download.

When working with label manufacturers, be sure to consider the use environment. Will the label be exposed to rain or snow? Is it likely to be abraded? Will it be exposed to intense sunlight or high temperatures? Figure 9.9A shows a label that has been rendered useless by abrasion. Considering its location on the side of the tread mechanism of a front-end loader (see Figure 9.9B), the label becoming scraped and torn is to be expected. In other words, it was reasonably foreseeable to the manufacturer that the label would be subject to abrasion. If a products liability lawsuit alleged someone was injured as a result of not having access to the information on that label,

FIGURE 9.9A Warning label rendered useless by abrasion. Labels must be designed, manu-factured, and placed to withstand the expected conditions of use for the product.

Source: Photograph by author.

FIGURE 9.9B In this instance, a different location or different kind of warning "label" (that is, more permanent than a stick-on label) would have been a better choice.

Source: Photograph by author.

the manufacturer could be held liable—even though the warning itself might have been flawless.

You probably will need to contract with a company that specializes in safety labels if your products are used in challenging environments. Local printers are generalists. They are skilled and competent at a wide range of printing needs. But they aren't specialists, and warning labels represent a printing specialty, not least because of the durability requirements. A durable safety label may need to incorporate a specialized base material and be covered with a clear layer to protect against abrasion. The adhesives, the inks, and the layers all have to be compatible with each other and the surface to which the label is applied—and be designed to withstand expected environmental exposures. Your local printer is probably not equipped to handle these requirements.

The good news is that better materials are being developed every day and there are companies that specialize in making state-of-the-art safety labels. Be cautious, however, that you choose one that really does have expertise in this area. Many companies market warning labels as part of their product line, but relatively few have the range of knowledge to ensure that your labels both meet standards and stand up to the environment.[27]

It may also be that a typical adhesive-type label is not the best choice for your product. Other options that may be used for warnings include enameled metal plates, hang tags, stamped or embossed warnings, and even woven cloth warning labels.

If you are concerned about the possibility that your warnings might deteriorate over time because of the use environment, make sure that your sales reps and service technicians are equipped with spare labels—if they encounter a machine with a missing or damaged label, they can replace it right on the spot. Also make sure that you offer replacement labels in parts catalogs. Whether you charge customers for replacement labels is up to you—but keep in mind that a 50-cent label might save you several thousand dollars in the long run.

SIZE AND PLACEMENT

How big should a warning label be? Where should you put it on the product? While there are no hard-and-fast rules addressing these questions, there are principles you can follow that will help you decide the appropriate answers. If you keep the goal of protecting the user at the top of your priority list, you'll be positioned to make the best decision.

Size of the label itself is not the critical variable: ability to see and read it—in time—is what's critical. If the text of the word message is in 6-point type and the pictorial (if any) is too small to decipher, the label will be essentially useless, regardless of its size. Some regulations specify a minimum letter height. For example, the regulations applying the *Federal Hazardous Substances Act*[28] specify minimum letter heights based on the size of the "principal display panel" of a label. So, for example, for a can of paint thinner, in which the required hazard warning is typically made part of the regular label on the can, the minimum letter size depends on the size of the can—and therefore the size of the label. Under those regulations, the smallest letter height permitted for signal word and state-ment of principal hazard is 3/64-inch, which is quite small. Regulations aside, that size type would be unusable for an on-product label for, say, a tractor or a punch press.

The ANSI Z535.4 standards tie recommended letter heights to "safe viewing dis-tance," which allows the standard to be applicable to a wide variety of products. The idea is that the text on a warning label should be large enough that the user doesn't need to get dangerously close to the hazard to read the label—which would rather defeat the purpose of having a label in the first place.

Placement of the label, like size, requires that you balance various factors to find the optimum location. The ideal location for an on-product label is one that ensures:

- The user will readily see it and be able to read it.
- The label will be far enough away from the hazard to view safely.
- The label will be close enough to the hazard to alert users to the danger.
- The label will be protected from damage.
- The label will not detract from product aesthetics.

Rarely will you find all these conditions. As with so much else in designing manuals and warnings, you will have to make trade-offs. If you keep the goal of protecting the user as the top criterion in making your choices, you will not go wrong.

Once you have decided on the appropriate placement for on-product labels, you may want to include an illustration in the manual to show where they should go, especially if it is foreseeable that the labels may have to be replaced due to wear. However, you should not rely on the user to apply the labels initially. For example, I once purchased an assemble-it-yourself TV stand that included a loose warning label to be applied to the top surface of the TV stand. What percentage of buyers of that product would be likely to slap an orange, black, and white warning label on top of a brand-new faux-walnut TV stand destined for the living room? Despite the fact that the label would be covered up by the TV, the percentage complying would probably be pretty low—perhaps about the same as would install anti-tip devices on their new kitchen range!

While the ANSI Z535 series comprise the only U.S. standards that address hazard communication for a wide range of products, many other standards that address specific products incorporate warning language and/or pictorials. Often, this language is compatible with the ANSI Z535 format. Sometimes, however, standards and even regulations conflict. A good example is the warning label shown in Figure 9.1, required by federal regulation. It does not conform to the ANSI Z535.4 standard at all. For example, the signal word "DANGER" is to be in red letters outlined in black against a white background—the reverse of the ANSI color scheme. If you manufacture lawn mowers, you have no choice but to follow the regulation, because if you don't, you can't legally sell your products. Regulations carrying the force of law always trump voluntary consensus stan-dards or industry "best practices." Remember, however, that you generally can supplement required labels with additional ones if that would be useful to help keep your user safe.

TESTING YOUR WARNINGS

The title of this chapter is "Warnings That Work." How can you tell whether your warnings are effective? One way, certainly, is to monitor your product's injury history. If you develop new warnings and the number of injuries declines, that might indicate that the warnings are effective—but the decline might also be due to factors unrelated to the warnings, such as improved training or work policies. In a way, the situation is similar to a decline in the crime rate after a new police chief takes office. The chief is happy to claim the credit, but the lowered crime rate might just as easily be related to an improving economy or a decrease in the number of 18- to 24-year-old males, the demographic group that statistically commits the most crime. The best way to evaluate the effectiveness of your warnings is by testing them, and doing so has additional benefits.

Why Test Warnings?

This chapter has discussed two goals in developing warnings: to protect the user and to protect the company against liability. Testing your warnings can help you accomplish both goals. If you test even before the product is put on the market, you have the assurance that your warnings are likely to prevent injuries, which, of course, in turn reduces your potential liability. Remember, if there's no injury, there's no liability. Rarely, however, will the injury rate be zero—even the best warning can be ignored. Testing your warnings will allow you to go to court prepared to defend them with confidence because your test results have shown them to be effective. The converse is true as well—if you don't test your warnings, you don't know whether they are effective. That means that you don't know whether you are doing all you can to protect your users, and you will have no comfortable answer when the plaintiff's attorney asks you, "Did you test this warning to find out whether it was effective?"

Even if you do test your warnings, if there is an injury and a lawsuit, be prepared for the plaintiff's attorney to attack your test protocol in an effort to prove that the test was inadequate. Testing is important, but a bad test may be worse than no test at all. Be sure you can defend your test protocol. Even if you're not an expert in testing and evaluation, you can develop and administer a defensible test, providing you follow a few principles of good test design.

Designing an Effective Test

The beginning of this chapter posed the question, "What is an effective warning?" An equally important question is "What is an effective test?" Without getting too deep into the science of evaluation, the short answer is that a good warnings test will give you useful information about whether the warning is effective. To do that, the test must be both valid and reliable.

Validity

Validity in testing refers to whether the test actually measures what you want it to. If you showed people three different warning labels and asked them to choose the best one, that would probably not be a valid test—because you don't know what criteria they are using to judge which one is "best." They might be choosing based on color preference, or how attractive the label was, rather than on how well it communicated hazard information. To ensure that your test is valid, you must decide exactly what you want to measure. In the case of a warning label, you would probably want to measure whether the pictorial and the word message (assuming you have both) accurately convey the nature and seriousness of the hazard, how to avoid the hazard, and what the consequences would be of not avoiding the hazard. That would require that the test subjects give a much more specific response than simply choosing the "best" one.

A valid test also tests only one thing at a time. If you show your test subjects a pictorial and word message together when trying to find out if the hazard information is communicated clearly, you don't know whether the pictorial and message each convey the information independently, or whether they can only do so in combination. If they don't work independently, then the pictorial may not be effective for users who cannot read. A much better approach is to test the pictorial and the word message separately. That way, you can gauge each one's effectiveness without the influence of the other.

Reliability

The other important characteristic of a good test is reliability. A reliable test ensures that similar performance produces similar results. In other words, when you test a given pictorial across different groups of people, the test should accurately tell you whenever someone does—or does not—interpret it properly. Good test design is critical to successfully evaluating the effectiveness of your warnings, but it is not the only factor you need to consider.

ELEMENTS OF A TEST

We think of a test as a piece of paper with multiple-choice questions on it or a practical skills evaluation like a driving test. The reality is that the test itself is just one piece of the process. A test actually consists of four separate aspects:

- The subjects taking the test
- The test itself
- The way in which the test is administered
- The scoring procedure and criteria

All four have to work together properly for the test to provide useful results.

Choosing Your Subjects

If you are going to evaluate the effectiveness of a warning, you need to make sure that you test it on people who are similar to your users. Just as the design engineer is not the best person to judge the effectiveness of a set of operating instructions (he or she already knows how to use the product and therefore might not notice ambiguous or misleading instructions), in-house "experts" are not the best choices for test subjects. Ideally, you should test your warnings on people who are or will be actual product users. Depending on your product, such people may be easy or difficult to find. For example, if you manufacture electrical equipment used primarily by electricians, it should be fairly easy to find a sampling of licensed electricians in your area. However, if the equipment might also be used by weekend do-it-yourselfers, be sure to include some non-professionals in your test group. Even if your product is too expensive or large for the average handyman—or woman—to own, if it is available on the rental market, you should include non-sophisticated users in your test groups.

If your product is a consumer product, for which the user group is highly variable, be sure to choose subjects that reflect that variety. Just because you have a college next door to your company, don't be tempted to recruit a group of college students as your test subjects. While they might represent part of your expected user group, they probably would not be representative of the entire group, especially in age and education level. The goal is to have your test subjects match as nearly as possible the characteristics of your actual user group. The marketing department should be able to help you find test subjects, since marketers typically do a good deal of research to pinpoint the demographics of potential customers. They may even run focus groups to help guide product marketing efforts—the same people who are useful in focus groups might also be useful to test warnings.

As part of your documentation, you should keep track of basic demographic and product-related information on each subject. Some of the information you would want to record, either from an interview or from the subjects filling out a form, would be the following:

- Age
- Sex
- Education level
- Occupation
- Training received relevant to this product (e.g., licensing, certification in a related field)
- Experience with this or similar products (what kind? how long?)

How many subjects do you need? The answer depends on what you need to show. If you want to prove with a statistical confidence level of 99% that your warning is effective, you will need a lot more subjects (and a professionally designed and administered test!) than if you simply want to get a sense of whether the warning is reasonably effective at communicating the necessary information. More subjects, assuming they are representative of your user group, are better than fewer. On the other hand, if you show a given pictorial to 25 people and every single one of them correctly identifies the hazard that it depicts, you may decide that's sufficient—nothing would be gained by testing another 25. But if three of your 25 subjects misinterpreted the pictorial, you might want to expand the test group to see whether those three misinterpretations were flukes or whether 12% of test subjects consistently misunderstood the pictorial.

Designing the Test

As discussed earlier, the test needs to be valid and reliable, but what's the best format? Should you offer a multiple-choice test? Short-answer? Should you ask closed-end or open-end questions? The advantage of the multiple-choice format is that it is very easy to score, which is why it is used for so many standardized tests administered to large groups of people. The disadvantage is that a multiple-choice test may not give you accurate information about whether your warnings communicate effectively. With the typical four-answer multiple-choice test, purely random choices will give correct answers 25% of the time. The more serious problem, however, is that multiple-choice tests limit the responses to the choices provided. That means that you will not get any information about possible erroneous interpretations of warning messages or pictorials that aren't among the choices. And you may get people choosing the correct answer by default, rather than because it actually matches what they understood the warning to mean.

A better option is simply to show the pictorial or word message and ask "What do you think this means?" Then allow the subjects to explain it in their own words. You may find ambiguities—especially in pictorials—that you would never have imagined, and thus would never have included in an array of multiple-choice answers.

If you have developed two or more versions of pictorials or word messages, test each of them—but not against each other with the same subjects. Give one group of subjects Pictorial A and ask its meaning, and give a different group Pictorial B, showing the same hazard. If 85% of test subjects correctly identified the hazard as depicted in Pictorial A, but 95% correctly identified it from Pictorial B, you know that B does a better job. Just like showing a selection of labels and asking which is "best," showing more than one version of a pictorial or word message might measure something other than its ability to communicate effectively.

Once you have identified a satisfactory pictorial and word message, then test the label as a whole, to make sure that the combination is at least as effective as each element singly. At this point, you can also ask more pointed questions such as these:

- What hazard is this label warning about?
- How serious is the hazard?
- What should you do to avoid it?
- What injury could happen if you didn't take action to avoid it?
- How likely would you be to get injured if you didn't take action to avoid the hazard?

The answers should give you a good indication of whether you need to revise the label.

Administering the Test

For your testing to be valid and reliable, you need to decide ahead of time exactly how you will administer it. Will you test subjects individually or in groups? Will you record oral responses or ask subjects to write their responses? (If you test in groups, you should use written responses to avoid having one person's response influence another's.) Will you show them the element being tested for a set time and then remove it, or will you leave it available throughout the procedure? (Showing the label for a set amount of time and then removing it makes scores more comparable and more meaningful. A label that takes five minutes to figure out would not be effective in the field.) Will you have the product available or test the label elements without reference to the product? If the product is available, will you let subjects handle it or just look at it? Whatever you decide, you need to make sure that you follow exactly the same procedure each time. That way you know that if the results are different from one test period to the next, the difference isn't a result of some change in procedure.

. You can use testing procedures to identify good placement locations for your warning labels or to find the most conspicuous place to put a warning on a product label as well. The basic principles are the same: test one thing at a time and administer the test the same way each time. Here's an example of how you could test label placement. Identify several different possible locations for a label on your product. Pick one to start with. Expose test subjects to the product for a limited time (e.g., a few minutes), then take it away. Ask the subjects:

1. Was there a warning label on the product?
2. Where was it located?
3. What did it warn about?

Then take a different group and do the same test, only with the label in a different location. You may find that the most conspicuous location is not the best for communicating information about a specific hazard—putting the label closer to the hazard (but not too close) may provide needed context for interpreting and remembering the warning.

Scoring the Test—How Good Is Good Enough?

This question has no easy answer. Naturally, we would like all warnings to be 100% effective at communicating hazard and avoidance information. The reality is that perfect communication is not always an achievable goal. It will be more difficult to achieve a high score with diverse user groups, such as those that buy consumer goods, than for a more homogeneous group. Groups that fall into the "sophisticated user" category, such as licensed or certified professionals with common training experience, make communication easier.

Where you set your "acceptable" score should depend also on how serious and likely the hazard is. If a person interacting with a given hazard faces almost certain death, you would want to require a near 100% success rate in communicating that hazard. For a lesser hazard, especially for a diverse user group, you might be satisfied with less. The absolute percentage is less important than your ability to show that you tried various options and chose the one that achieved the highest score. On the other hand, if you have a "DANGER"-level hazard (" . . . will result in death or serious injury"), and the best score you can achieve with a warning is 70%, it may be time to look at a redesign.

Testing your warnings, like conducting a hazard analysis, should not be a one-time exercise. Standards evolve over time; people become educated to certain hazards; literacy and graphic understanding in various populations increase or decrease. If your product goes through a significant redesign or you introduce new models with different features, you may wish to retest the warnings that you have been using for a while to make sure they still work. There is no hard-and-fast rule about how often you should retest, but certainly if there is a significant change in the product or in the population of users, you should consider testing the warnings again.

Your testing protocol and the results should be well documented. The same arguments for and against documentation of a hazard analysis apply to warnings testing—and you're still likely to be better off maintaining records than not. Doing so demonstrates a strong commitment to product safety and to your users' well-being.

While it is not difficult to set up a reasonable testing protocol in-house, another option is to contract with a firm that provides testing and evaluation services. Working with an outside firm has the advantage of ensuring an unbiased test and independent results. Just as you may send your product to an independent testing laboratory for certification that it meets acceptable safety standards, you may find it useful to have an outside review of your warnings.

SUMMARY

Managing the residual hazards in your product (those that cannot be designed out or guarded) generally requires developing warnings, both on-product warning labels and safety messages in manuals. An adequate warning must give the user the information needed to make an intelligent choice about whether to use the product at all, and, if so, how to use it safely. To be considered adequate, generally a warning must inform the user of the nature and severity of the hazard, tell how to avoid the hazard (unless these are obvious), and communicate clearly. Clear communication is usually the most difficult element to achieve. In general, you will be more likely to communicate clearly if you focus on keeping the user safe rather than protecting the company against liability.

Good warning labels typically include a signal word, pictorial, and word message. Choosing the correct signal word is essential to ensuring that the warning is adequate. The signal word alone should accurately convey the seriousness and likelihood of injury. The pictorial (if used) and word message are left to explain the nature of the hazard and how to avoid it. Pictorials serve to reinforce the word message for most users. For users who cannot read the word message, pictorials alone must convey essential hazard information. To be effective, pictorials must be simple and easily understood. Similarly, the word message must use simple, direct language and be specific enough to be helpful. Finally, good warning labels must be durable under the expected conditions of product use.

To be considered adequate, warnings must also meet applicable standards. These vary from product to product, and range from government regulations to voluntary industry standards. Some products may have no associated standards at all.

Three federal regulations that apply to a wide variety of products are the *Federal Hazardous Substances Act,* the *OSHA Hazard Communication Standard,* and the *Consumer Product Safety Act.* Many other federal regulations apply to particular product families, such as pesticides or children's toys. Voluntary standards are often developed by industry groups, such as trade associations. While these are technically voluntary, in practical terms they may be considered the minimum that companies must meet.

Two standards for warning labels in wide use are ANSI Z535.4 *For Product Safety Signs and Labels* and ISO 3864–2 *Graphical symbols—Safety colours and safety signs, Part 2: Design principles for product safety labels.* In an increasingly global marketplace, deciding which of these would be more effective for communicating hazard information can be difficult. Fortunately, these two standards-setting organizations have worked diligently toward harmonizing their formats, making it easier for manufacturers to achieve standards compliance both in the U.S. and abroad. An additional standard in ANSI's Z535 series is ANSI Z535.6, *Product Safety Information in Product Manuals, Instructions, and Other Collateral Materials,* which provides guidance for consistent formatting of safety information in manuals.

To be sure that your warnings are effective, you should test them. A defensible and useful test must be both valid and reliable and be relatively easy to implement. Fortunately, even small companies can design a simple yet effective testing protocol. Key elements include choosing test subjects similar to expected users, testing one thing at a time, and ensuring that the results are comparable by following strict procedural guidelines. Testing your warnings gives you valuable information about their effectiveness and also demonstrates a concern for user safety.

Developing effective warnings, like writing good manuals, requires the ability to communicate well within your company as well as with your customers.

CHECKLIST: MAKE YOUR WARNINGS WORK

- Have I identified residual hazards that need on-product labels?
- Am I using the correct signal word, based on severity and likelihood of injury?
- For each label on the product, is there a corresponding safety message in the manual? Does it use the same signal word?
- Are my pictorials immediately clear and easy to understand?
- Is the word message simple, direct, and specific?
- Will my labels be durable under expected conditions of use?
- Are stick-on labels the best choice? Would another approach work better?
- Are the labels big enough to read easily at a safe distance?
- Have I met the requirements for warning language in applicable government regulations?
- Do my warnings meet industry standards?
- Do my warnings comply with ANSI Z535.4, ANSI Z535.6 and/or ISO 3864–2 as appropriate for my market?
- Have the warnings been tested?

NOTES

1. Thomas J. Ayres, Madeleine M. Gross, Christine T. Wood, Donald P. Horst, Roman R. Beyer, and J. Neil Robinson, "What Is a Warning and When Will It Work?" in *Human Factors Perspectives on Warnings*, ed. Kenneth R. Laughery Sr., Michael S. Wogalter, and Stephen L. Young (Santa Monica, CA: Human Factors and Ergonomics Society, 1994), pp. 1–5.

2. *Restatement of the Law Third, Torts: Products Liability* (American Law Institute Publishers, 1998), § 2, i.

3. From Material Safety Data Sheet for Pirtek Hydraulic Oil 10, www.pirtek.co.uk/downloads/Pirtek%20Hydraulic%20Oil%2010%20-%20SDS22742.pdf, accessed May 1, 2019.

4. See, for example, the Safety Data Sheet for Liquid Drano®, https://corp-ucl.azureedge.net/-/media/sc-johnson/our-products/sds/us-english/home-cleaning/350000004298-drano-liquid-drain-cleaner-08-16-2017-2-0-en.pdf, accessed May 1, 2019.

5. See, for example, Kenneth R. Laughery and Julie A. Stanish, "Effects of Warning Explicitness on Product Perceptions," in *Human Factors Perspectives on Warnings*, ed. Kenneth R. Laughery Sr., Michael S. Wogalter, and Stephen L. Young (Santa Monica, CA: Human Factors and Ergonomics Society, 1994), pp. 125–129; and Kent P. Vaubel and John W. Brelsford, "Product Evaluation and Injury Assessments as Related to Preferences for Explicitness of Warnings," *Proceedings of the 1991 Human Factors and Ergonomics Society Annual Meeting* (1991), 1048–1052. See also Kenneth R. Laughery, Sr. and Danielle P. Smith, "Explicit Information in Warnings," in *Handbook of Warnings*, ed. Michael S. Wogalter (Mahwah, NJ: Lawrence Erlbaum Associates, 2006), pp. 419–428.

6. Even if your state allocates liability so that the user may be held partially liable for an injury, disclaimer-type warnings generally will not shift the burden to the user more than the facts of the case would on their own merit.

7. 15 *U.S. Code* 1261.

8. 29 *Code of Federal Regulations*, Part 1910.1200.

9. 15 *U.S. Code* 2051–2084

10. Note that this explanation is necessarily quite simplified.

11. 16 *Code of Federal Regulations*, Part 1500.121.

12. 29 *Code of Federal Regulations*, Part 1910.1200(a)(1).

13. See, for example, Benoit Nemery and Jerrold L. Abraham, "Hard Metal Lung Disease: Still Hard to Understand," *American Journal of Respiratory and Critical Care Medicine* 176 (2007): 2–3.

14. Again, this is highly simplified. For a full explanation, consult the standard.

15. See OSHA Brief, *Hazard Communication Standard: Safety Data Sheets*, www.osha.gov/Publications/OSHA3514.html, accessed May 9, 2019.

16. 16 *Code of Federal Regulations*, Chapter II, Part 1205.6.

17. 16 *Code of Federal Regulations*, Chapter II, Part 1211.17.

18. 15 *U.S. Code* 2056 §7(b)(1).

19. *ANSI Z535.4–20011 (R.2017) For Product Safety Signs and Labels* (Rosslyn, VA: National Electric Manufacturers Association, 2017). For a quick look at the development of the ANSI Z535 standards, see, Geoffrey Peckham, "A Brief History of the ANSI Z535 Standards," *NEMA Electroindustry*, February, 2016, p. 41, www.nema.org/Communications/EI/Documents/Peckham-clarion-Z535.pdf, accessed May 1, 2019.

20. *Product Safety Sign and Label System* (Santa Clara, CA: FMC Corporation, 1985).

21. *Product Safety Label Handbook* (Pittsburgh, PA: Westinghouse Electric Corporation, 1981).

22. *Globally Harmonized System of Classification and Labeling of Chemicals (GHS)*, 2nd rev. ed. (New York and Geneva: United Nations, 2007).

23. "Symbol Making," *New York Times Magazine*, November 18, 2001. In fact, one of the reasons it was picked was that when tested on consumers, the symbol had the fewest associations of any that the examiners tested. The symbol was developed at the Dow Chemical Company during the 1960s. As one of the developers, Charles Baldwin, put it, "We wanted something that was memorable but meaningless, so we could educate people as to what it means."

24. It's all right to use "recommended" as a qualifier, as in "Keep tires inflated to recommended pressures."

25. ISO 3864–2, *Graphical Symbols—Safety Colours and Safety Signs—Part 2: Design Principles for Product Safety Labels* (Geneva, Switzerland: ISO, 2016).

26. Sometimes, the black border is surrounded by a yellow border.

27. One that does have expertise on both counts and has long been a leader in the field is Clarion Safety Systems (formerly Hazard Communication Systems), Milford, PA. Their website also has useful information about standards and other resources: www. clarionsafety.com.

28. The *Federal Hazardous Substances Act* is 15 *U.S. Code* 1261; minimum letter heights for labels to conform with the act are set forth in 16 *Code of Federal Regulations,* 1500.19(d)(7).

Part IV

Making It Work in the Real World

10 Who Writes a Manual?

OVERVIEW

So far, this book has focused on the user who reads the manual and on the manual itself, what it should look like, what it should include, and how it should read. But what about the person or team of people who creates the manual? Who are technical writers and how do you become one? Why would you want to? And where do you fit into the company? This chapter looks behind the scenes at the business of making manuals. Note that while this chapter is focused on strategies for working within a company to develop a traditional print manual, most of the same strategies apply also to more cutting-edge ways of delivering product content—videos, interactive animations, on-demand information, etc.

Since the manual serves as an ambassador for the company, it makes sense to choose its writer or writers well. As we will see, the role of the technical writer has changed over the years, along with the typical entry routes and training options to become part of the profession. One thing has not changed—even if you are the only technical writer in your company, you are still working as a member of a team. Producing a manual requires input and assistance from many quarters, and the most effective writers are the ones who develop the best networks. Networks also help to overcome functional divisions within the company. Technical Publications (Tech Pubs) is seldom a stand-alone department, and being able to work across organizational chart lines regardless of where in the company you're located is a critical skill.

Writing manuals is difficult enough; managing documentation projects is even harder. The challenge is always to do more with less and somehow make sure that the moment the product is ready to ship, the manual is ready to go along with it. This chapter provides some tools to help you work within the constraints of your organizational setting and resource limits and still produce top-notch manuals. If that seems like a tall order, it is; but there's never been a better time to be in technical publications.

TECHNICAL WRITERS—LOW ON THE FOOD CHAIN NO MORE

Remember Rodney Dangerfield? He's the comedian who didn't "get no respect," no matter what he did. Well, it used to seem that the same was true of technical writers. Forty years ago, you could not major in technical writing in college, and very little was published about how to write manuals because nobody thought they were all that important. In fact, the task of writing manuals was often given to engineers who weren't very good at engineering. Writing manuals was a low-status position, and in many companies, a dead-end job. Now many colleges and universities have technical writing programs, the median salary for technical writers as of this writing is over $71,000 per year, and technical writing has become a respected profession with its own professional associations and journals. Successful technical writers often have many opportunities to advance in their careers, including moving into senior management positions.

What changed? Several factors contributed to the much-improved position of technical writers. One of these was the products liability lawsuit. As we have seen, poor instructions and warnings can be considered a product defect. When plaintiffs' attorneys began to attack badly done product manuals, manufacturers began to pay more attention to producing good ones. At the same time, manufacturers started to see the customer relations value in turning out top-quality manuals.

THE VOICE OF THE COMPANY

In a real way, the technical writer is the voice of the company. It is, after all, the technical writer's choice of language, style, and format that convey a sense of the company to the customer. As noted in Chapter 1, the manual is an important part of the company's brand, and the technical writer is in a sense an ambassador for the company. It stands to reason that a company would want to put its best foot forward, so to speak, and ensure that its technical writing staff was comprised of well-trained professionals.

Training for Writers

Writers need training for manual production, just as they do for any other kind of specialized publication. While some companies provide no training whatsoever or give their writers only the sketchiest of orientations, others have systematic and comprehensive training programs. Generally, companies that have been producing products for a long time, especially if they are also large companies, often have sophisticated and comprehensive training programs for manual writers. If you work for such a company, you may already have had orientation sessions; hands-on practice working with other, more experienced writers; and close contact with your publication managers and editors. If your company currently has no such training program, you (or your manager) might consider some of the following training techniques:

- Have new writers work through a manual from start to finish with an experienced writer.
- Give new writers in-house handbooks and style guidelines or workbooks to orient them to company procedures.
- Have managing editors work closely with new personnel in the first months on the job.
- When companies are decentralized and manuals are produced at several places, assign one manager to coordinate quality control of the manuals. (Some companies use a "roving editor" who travels among the various manual production locations.)
- Bring writers together for periodic training sessions on such special work topics as international standards, writing safety warnings, graphics techniques, or integrated product safety (see Chapter 11).
- Make libraries and files of company and competitor manuals available for writers to look at.

- Teach writers "incrementally" by assigning only small segments of a manual for their first assignment, so that they gradually expand their skills and techniques.

If your company is too small or has too few writers to make in-house training sessions cost-effective, you can consider using periodic outside consulting help. Many smaller companies make use of the continuing education conferences and seminars conducted by universities and technical institutes. Working with consultants and attending continuing education seminars gives writers a chance to learn and exchange ideas with other professionals, to bring themselves up to date on products liability, and to practice their writing skills.

Other Helps for Writers

If your company makes no provision for organized training sessions, you can still ease the writing process a great deal by creating style handbooks, writer guidelines, and fact sheets listing steps involved in the manual process. Managers or editors can provide such books for their writers, and solo writers can create their own handbooks to systematize the procedures they plan to use. For example, one company's fact sheet, given to writers before work begins on the manual, contains the following information:

- Product name and number
- Deadline dates for completed manuscript in rough form
- Format specifications (column and page width, margins, type size, specifications for photos and drawings)
- Schedule and locations for viewing the product and for hands-on practice with mechanisms of the product
- Style guidelines (average sentence length, vocabulary and language level, use of active-voice verbs, etc.)
- Notifications of what the other segments of the manual will be and which writer is assigned to those segments (especially important for cross-reference work or for machine systems that interact)
- List of names and contact information for people who can provide information
- Lists of key meetings for product development
- References to materials on file that might be reused

In addition to providing writing help, pay attention to the influence of the physical workplace. Noticeable improvements in writers' effectiveness may come with attention to such mundane details as adequate lighting, up-to-date workstations, and ergonomic chairs. For example, one company found that the time to produce a manual could be cut significantly simply by relocating the art department to the same floor as Tech Pubs!

Which Is Better—Engineer or English Major?

When a company decides to assign the manual-writing task or to hire a new writer, the question is often posed, "Should we choose technicians and engineers and then teach them how to write, or should we choose professional writers and teach them the technology of the product?" Stating the question this way can be misleading because of the underlying assumptions—that technicians and engineers can't (or won't) write well and that trained writers will probably be technically naïve or ignorant.

A better way to think about the choice is to pick someone who can communicate and likes to. Certainly, some lawyers and engineers are so bound up in jargon that they find it almost impossible to simplify a message for a general public audience, and some wordsmiths write clean and explicit prose, yet are so technically inept that they cannot grasp the workings of the simplest machine. Nevertheless, there are individuals who possess a combination of the necessary communication skills and technical ease, and these make the best technical writers. These people recognize clear, correct prose and can also write it. They have a visual sense about drawings, photos, and format devices, and have good "people" skills. They also are interested in how things work and either know the product from experience elsewhere in the company, or can, with a minimum of explanation and hands-on practice, quickly grasp a new technology or product. Curiosity about the technology is often every bit as important as technical training. A curious technical writer will ask questions— and often reveal gaps in existing documentation that no one else noticed because of "shop-blindness."

As part of the hiring process, it's a good idea to devise some simple screening devices to identify good communicators. For example, you could give applicants a simple device or product, along with basic information and relevant photos or drawings. Then ask them to write and lay out a sample page of a manual. Another method is to give applicants sample pages of existing product manuals from a variety of products. Choose a spectrum of quality (some good, some bad, some average) and ask applicants to rank the pages and to justify, in writing, the reasons for their choices. Both of these methods should give you a pretty good idea of which applicants will build good relationships for the company by developing high-quality manuals.

When I Grow Up, I Want to Be . . . a *Technical Writer*?

Somehow, technical writer never seems to rank up there on kids' wish lists with police officer, firefighter, astronaut, doctor, lawyer, teacher, and so on. That is beginning to change, but even now, if you poll a group of technical writers as to how they came to their profession, you are apt to get a great variety of responses—with most of them suggesting an accidental (or serendipitous!) career path rather than a planned one. Here are some of the responses from a poll of attendees at a recent technical writing seminar about their backgrounds:

- French literature major
- Engineering major
- Graphic artist
- English major
- Company owner
- Marketing major
- Science writer

While writing manuals is often a profession that one "falls into," it is increasingly possible to obtain a college degree that is specifically geared toward preparing one for a career in technical communication. These range from a four-year bachelor of science (BS) or bachelor of arts (BA) degree, to two-year associate's degree (typically an associate of applied science or AAS degree), to a certificate program, which is a concentration of courses within another degree program. Some of these programs are offered in an online or other distance-education format, making them accessible to anyone with an Internet connection.

Here are a few examples of colleges and universities offering technical writing program:

- Carnegie-Mellon University: Four-year BS degree in technical writing
- University of Minnesota: Four-year BS technical writing and communication
- Arizona State University Online: BS in technical communication
- Fox Valley Technical College: Two-year AAS degree in professional communications
- University of Wisconsin-Stout: MS in Technical and Professional Communication

Just as academic programs and degrees have proliferated, technical writing has developed as a profession in its own right, with conferences, seminars, professional societies, newsletters, and books. The Society for Technical Communication, for example, is a national organization with regional and local chapters. Its publications and conferences help writers stay current in the field and keep in touch with each other.

Increasingly, as the technology both for products and for communication media become more complex and specialized, the requirements for even entry-level technical writers will also increase. The days of slap-dash manuals turned out by out-of-work English literature majors or whichever engineer or technician was lowest in the pecking order are long gone. Today's technical communicators are—and need to be—trained professionals.

IF WE'RE SO SMART, WHY AREN'T WE RICH?

The good news is that technical writers are getting richer . . . well, at least better paid. The latest salary information from the Department of Labor puts the median annual earnings for technical writers at almost $72,000—with the top 10% earning more than $114,930.[1] While that doesn't put a technical writer in the same league as Bill Gates, it's a grand improvement over the salary picture just a few years ago. The demand for technical writers is expected to grow at or above the average rate for other occupations as well—a projected 11% increase by 2026. Industries are beginning to recognize that manual writers are the bridge builders between the product and the consumer.

As products grow more complex, as formerly simple mechanical devices are steadily being electronically controlled and computerized, and as manuals are interpreted by courts as significant legal evidence, the technical writer's work is becoming more valuable—and more highly valued. The new attitudes are reflected in better salaries, more investment in writing training programs, and better integration of the technical writer into the mainstream of company organizational structures.

WRITING MANUALS IS A TEAM SPORT

The stereotypical image of a writer is the morose and solitary figure laboring over a manuscript in a garret apartment (preferably in Paris), who engages in conversation only in the evening while drinking wine in a sidewalk café. Does that sound like your job? Probably not—except perhaps for producing the urge to drink! In all companies, regardless of size, writing manuals is a team sport, not a solo pursuit.

HOW MANY WRITERS DOES IT TAKE TO MAKE A MANUAL?

Whether the technical writing "department" is made up of one or many writers, producing manuals is still a team sport. A solo technical writer cannot make a manual without the help of the engineers, marketing people, and service people who provide necessary information and feedback. Even so, solo and team writing do differ in important ways. Naturally, each situation has its own pros and cons.

The Solo Writer

Small companies with just a few products may assign one person to do the manuals for all of them. If you are that solo writer, you will soon find that writing is only one part of your job. You may also have to take photographs (or scan them), plan the artwork, choose paper stock, edit, type, and even print the entire manual. You will want to clarify, before you begin, how much control you have over the choices you'll have to make. Ideally, as a solo writer the control over many of these decisions should be yours. If people you talk with seem vague about who decides questions about layout, organization, and so on, try to claim the authority and the responsibility for the technical writing decisions. Convince others that you are the resident expert.

As a solo writer, you have many opportunities to be creative. Because the majority of decisions will fall to you alone, you can approach the manual production job with your own vision of how the final manual will look, and you can make certain decisions without having to clear each step of the production with someone else. Many solo writers say that the autonomy they enjoy more than compensates for their many responsibilities. They also value the variety of tasks involved and enjoy the different kinds of people they work with. Most of all, they like having control over the project from start to finish.

On the other hand, if you are a solo writer, your work will be the single bridge between the technical data about your product and the manual that reaches users. You will have to gather the information and create the schedule yourself. Manual writers who work solo often feel rushed, isolated, and pressured by their many responsibilities. They sometimes feel that other people, on whom they must rely for information, have little understanding of, or sympathy for, what it takes to put a manual together. In particular, they think other people often believe manual writing "happens" faster than it does, that it's a quick process. It's not, of course. But others may not understand this fact of manuals specifically, or of writing in general.

How can you make solo writing easier? You have the disadvantage of not having a writing team to lean on, either for moral support or advice. You don't have comrades in the writing trenches who understand how difficult your job can be. The solution in part may be to educate your co-workers. For instance, you might begin by simply keeping a record of how many hours it usually takes you to create a manual page for a new product or how much time it takes to produce a computer-generated graphic. Even relatively uncomplicated record keeping can be a big help if you have some facts and figures to show that deadlines are unreasonable, costs are too high, or one person can't do the job alone.

Your needs as a solo writer are much the same as the needs of team writers. You need access to information and time to do the job. As you read the rest of the chapter on team writing and on information and time, you will find many suggestions that you can adapt to the solo-writing setting. Consider, especially, ways in which you can perform the same functions as a team leader performs in team writing. For example, you can do your own planning by:

- Arranging your own schedule of meetings with key personnel to collect information
- Asking for help from informal support teams or individuals (for work such as typing, drawing, planning safety messages, taking photographs)
- Developing a thorough outline
- Laying out steps in manual production
- Setting up a style and format guide or handbook so that your own writing procedures become standardized and easier to repeat from manual to manual

Team Writing

In larger companies, manual writing tasks may be spread over several writers, graphic artists, and editors. A team of people may be assigned to develop the manual for a single product. This may be especially likely if the product is complex. Such a division of labor makes sense for a number of reasons: preparation times can be shortened, writers can develop special expertise with certain manual segments, and teams can include personnel from other company units (e.g., technical, research, product safety). The team-written manual also poses problems, however, particularly those of conceptual unity, team coordination, and uniformity of quality. Here are some of the pros and cons of the team-written manual and some suggestions for making team writing efforts smoother.

"Many hands make light work" used to be a common saying. It's still true. Dividing the manual writing according to systems or processes inherent in the product or according to special areas of writer expertise allows you to make the best use of writer talent and to get the job done more quickly and accurately. For example, a writer whose specialty is electrical systems will be able to write about that aspect of a product more easily (and probably more quickly) than the jack-of-all-trades writer who has to keep many different kinds of processes or procedures in mind.

Situations also arise in which a machine or product that has used standard mechanical or chemical processes for years is then suddenly altered by new technology or by the addition of an electronic component. These kinds of alterations have frequently occurred, for example, in products involving computers, numerical control, and robotics (e.g., devices for welding, spraying, assembly procedures, systems integration, and quality control). In such cases, the best use of talent may be to ask the technician or engineer-designer who created the new component to write the segment describing its function. An editor can then review the segment to make sure the style is user-friendly and not filled with technical jargon.

"The camel is a horse designed by a committee" is another saying, and one that applies here as well. Too often, the team-written manual has camel-like lumps and bumps. Such manuals move by fits and starts from one segment to another. They sometimes have ill-matched writing styles and formats. Users find these manuals very hard to use because of their redundancy, lack of cross-referencing, and chaotic organization. In brief, the chief difficulty with the team-written manual is the coordination of several writers' work into a smooth manual that looks and sounds as if one person had written it. The team-written manual is a reflection of company structures and procedures as well as management styles and individual personalities. Coordination of team writing should make the best use of available time, talent, and money. Who decides what and when are questions that need to be answered up front.

Team-written manuals often have a team leader, a manager, or a service publications editor who has the final responsibility for, and a unified concept of, what the finished manual will look like. That unified concept may be the joint creation of the writing team, but once the conceptual framework is established, leaders are often responsible for the scheduling, assignment of manual segments, creation of clear instructions for what each manual segment will include, and final coordination and editing of the completed manual. Team leaders should have strong writing and editing skills because they will have the job of making language, style, and format internally consistent. A good team leader will make use of instructions, writer guidelines, writer checklists, regular (even if informal) meetings or check-ins—any procedure that helps writers know what is expected of them and when.

WORKING WITH GRAPHIC ARTISTS

Even if you are a one-man band, so to speak, as a solo writer you will probably need to be able to work with a graphic artist to develop illustrations for the manual. As Chapter 5 notes, simply reproducing engineering drawings for use in the manual is usually not the best choice. It's a much better idea to draw illustrations specifically designed for the manual. The first step is to figure out what illustrations you need.

Chapter 5 identifies a number of situations in which pictures usually work better than words. As you develop the outline for your manual and create your storyboards, if possible, involve the graphic artist. Your job is to convey to the artist exactly what you need the illustration to show. The artist's job is to figure out the best way to do that. Too often, artists are given vague instructions like, "I need a drawing of the lift arm." The result is a vague drawing that is not useful to the manual reader. Even if the artist asks you "what kind" of drawing you want, resist the temptation to answer with something like "a line drawing" or "an exploded view." What the artist really means is, "What's the purpose of the drawing?"; is it:

- To enable the reader to identify parts?
- To show the location of the lift arm with respect to the rest of the product?
- To indicate how it is to be assembled?
- To illustrate a range of motion or specific positioning?
- To show lubrication points?

Each of these different purposes might call for very different drawings.

Just as your writing becomes more focused (and useful) if you have a clearly defined purpose as well as a subject, so an artist's drawing will be more focused and helpful if he or she clearly understands what the illustration is intended to convey. The key to both is developing text and graphics together, as Chapter 5 suggests. You will be better able to do that if you take the time to learn a little of the artist's craft. This doesn't mean that you have to learn to draw; it means simply that you should spend some time with graphic artists and ask questions. Ask how they choose one type of illustration over another; ask questions that will help you understand their design process. The more you can learn about how artists think, the better you will be able to explain your needs in terms they understand.

TAKE AN ENGINEER TO LUNCH (OR: CULTIVATE A SOURCE)

Do you remember the instructions your high school English teacher gave you about how to write a paper? Put in the context of manuals, it probably went something like this:

1. Make all basic document design decisions
   Content
   Format.
   Schedule
2. Gather information
   About product
   About users
3. Prepare outline and list of visuals
4. Write entire draft and prepare all visuals
5. Edit and get approvals
6. Print manual
7. Distribute manual with product

Does that sound like your reality? Probably not. Your reality probably looks more like this:

1. Receive assignment with tight deadline.
2. Begin making some basic decisions.
3. Deadline is moved up two weeks.

4. Try to get information from engineers. Receive spec sheet with illegible handwritten changes. Receive torn copy of competitor's brochure.
5. Try to get product. Receive outdated model with parts missing.
6. Deadline is moved up two weeks.
7. Start to write anyway. Receive current prototype. Celebrate.
8. Overhear hallway conversation about radical design changes in product.
9. Scrap draft. Begin to read want ads. Consider the benefits of consuming large quantities of wine in a sidewalk café.

The inevitable constraints of a competitive economy mean that you will never have enough time to write the perfect manual—because the manual and the product should ideally be finished at exactly the same moment, so they can ship together on schedule. The equally inevitable reality is that the final product won't look exactly like the first prototype—and may look very different indeed. Like a writer on the seventh draft, engineers always want to tweak something to make the product just a little better.

How can you possibly get the information you need—on time—to do your job? If you have problems getting information, you're not alone. Many writers have said that their chief frustration (even more than tight schedules and low pay!) was a lack of information about the product. Engineers are out of town troubleshooting another installation. Technicians can't hook up the system because they're working on another project. Designers are too busy to explain the basics. Writers may ask to see the product and be refused. They may ask for scheduled time to review the product with designers, technicians, engineers, or safety personnel and be told that there is no time. They may ask for a working model, a prototype, or at least a photo and get a flat no for an answer—or they may be housed in an office miles away from where the product is designed and manufactured.

Admittedly, many people may be clamoring for a prototype of a new product. Marketing wants it, engineering is working on it, and product safety needs it. When the pressure is on and deadlines must be met, writers often come in last. However, if the manual is to perform its function, writers must have information and management must provide the procedures to help them obtain it. Information gathering is an important first (and ongoing) step in planning the manual. Unlike the tortured, would-be author in the Paris garret, technical writers spend much of the workday out and about, talking to people—engineers, technicians, service experts, even customers—because they have the information the writers need. However, these people are all busy with their own jobs. How can you get them to take time to help you?

In a very small company, the writer may need only to lean across the desk and ask a co-worker for information. In very large companies, an email might need to go halfway around the world. Successful writers often use three key tactics for coping with these obstacles:

• Becoming part of the product development team
• Cultivating key contacts
• Using modular design

Note the phrase "coping with these obstacles" rather than "solving these problems"—they will never be totally solved. These three tactics, however, will help reduce the chaos.

Become Part of the Product Development Team

If your company practices concurrent engineering, in which the manual and product are developed together, they'll understand this concept a little better. The earlier you can be involved in the product development effort, the earlier you can start writing the manual and the more timely information you will have. If your company does not already have the technical publications department as part of the product development team, volunteer. As one technical writer put it, "I just started going to meetings to which I wasn't invited. After a while, they expected me to be there!"

Also show up at hazard analysis/product safety meetings. Knowledge of safety concerns is vital, and your "outsider" position may help others who are more closely focused on creating the physical product see the design from the user's point of view.

Ideally, the manual and the product should be designed in sync. Try to make that happen in your company.

Cultivate Contacts in Key Areas (and Do a Few Favors)

Develop relationships with one or two people in engineering, service, and so on. If you cultivate a good working relationship with someone, that person will be more likely to pick up the phone and call you to tell you about a design change rather than to wait passively until you come to him or her. Developing these contacts may mean trading favors. As the writing expert, you may be asked to edit letters or look at documents that fall outside your job description. When you can, be friendly and helpful. You'll be in a better position to ask for help when you need it.

Use Modular Design

Chapter 3 suggests planning the manual in modular sections—because it makes it easy to update or create "custom" manuals for different product models. Another advantage is that you can work on draft modules independent of one another. Even if you don't have all the information for Section A, for example, you can still work on Section B. In addition, the modular writing technique makes it a lot easier to revise the manual to reflect design changes—whether those changes take place after the product has been on the market a while or before it's ever out the door.

Technical writing in the marketplace will never be as tidy as your high school English teacher envisioned. Writing manuals is like shooting at a moving target: products constantly change to incorporate innovation or meet changing market needs. The manuals that accompany those products must also change. Good technical writers simply find ways to minimize the disruption when the changes happen—again.

PLAYING WELL WITH OTHERS: MARKETING, LEGAL, DESIGN, MANUFACTURING

As a technical writer, you often wear many hats—writer, interviewer, customer advocate, and so on. Here are three more roles you often play within your own company:

- Traffic controller
- Translator
- Troubleshooter

You may be a traffic controller because, quite often, all information roads lead through the technical publications department. In many companies, this department is the only place where sales literature, service and user manuals, parts lists, training manuals, engineering specifications and computer-aided design (CAD) drawings, graphics, video, film, and computer information all come together, or at least pass through. On a good day, traffic on this information highway is heavy, but smooth. On a bad day, this publication intersection has only fast traffic and no stoplights—and no nearby exit ramps for overrun writers.

Technical writers often find themselves working as translators for the technologically impaired or technologically injured. Bruised by products that are beyond their sophistication, poorly designed, or aggressively user-unfriendly, consumers look to manual writers for help. Sometimes it works the other way—you have to explain to engineers or others how a customer will (or will not) understand the description they've written. Remember, calling the supports for a camping tent "aluminum beams and columns" may be crystal clear to an engineer, but the rest of the world calls them tent poles.

As a troubleshooter, you can be one of a company's unrecognized assets when it comes to product development. If brought into the process early enough—say, from the beginning—technical writers can provide a fresh, often non-technical yet technically friendly, questioning perspective. They can help with some design choices and safety analysis. Much like the software interaction designers mentioned earlier, technical writers can help engineers, programmers, and other research and development people see the user before they envision the product.

Because of these varying roles, technical writers usually find themselves at the hub of a wheel, with the spokes being other departments, including all these:

- Research and Development
- Legal
- Design
- Manufacturing
- Marketing
- Sales
- Service
- Financial

In fact, just about every other part of the company, with the possible exception of Human Resources, will interact with Tech Pubs on a regular basis. As a technical writer, your work product depends on inputs and sign-offs from these other areas of the company—but the reverse is not always true. Design engineers do not need input (or think they don't) from Tech Pubs to do their job. Sales personnel do not need information (or think they don't) from Tech Pubs to sell their products. But you do need information from them—so it's up to you to build relationships and win consensus. Making yourself useful as a translator and troubleshooter will certainly help, but ultimately your success may depend in part on where you're located in the company's organizational chart.

Location, Location, Location: Where Do You Put Tech Pubs?

If you survey technical writers to find out where their department is located within their company, you will find the vast majority are placed within one of three major divisions:

- Engineering
- Service
- Marketing

Occasionally, you will run across a writer attached to Legal or Safety, and very occasionally, you will find Tech Pubs as a stand-alone division. In some companies, of course, the documentation function is totally outsourced, and there is no in-house department, but this is rare. Outsourcing is used more often to handle temporarily high work flow.

Why is there no established place for Tech Pubs? The primary reason is that most manufacturers still imagine that their product is the printing press or bulldozer or dishwasher or rolling mill that emerges from the factory, when in fact the product is actually the machine *and* the manual *and* the marketing literature. Increasingly, the courts in the United States are starting to perceive the product as all those; in the European Union the manual is already explicitly identified as part of the product. But if the mindset is that "we make concrete saws," it's easy to see the need for an engineering division, a manufacturing division, and a service division. It's not so easy to see the need for a coequal technical publications division.

What's the best place for a technical publications unit? Good question—and if you haven't thought a lot about it, you're not alone. As Nina Wishbow points out in "Home Sweet Home: Where Do Technical Communication Departments Belong?"[2] academic studies have not explored the issue. Even Karen Schriver's encyclopedic *Dynamics in Document Design*[3] addresses the question mainly from the perspective of the profession at large. However, Schriver's wonderfully titled section "Practitioners without a Profession: 'Nobody Loves Me But My Mother and She Could Be Jivin' Too'" could easily describe the place of technical writers in most companies. Whether or not the question has been explored, the reality is that in writing, as in real estate, location does matter.

Different Roles Mean Different Agendas

We all tend to see the world through the lens of our own mission, and the perspective on technical publications is no different. Members of the engineering division will see the manual as a means to explain how the product works* and what its capabilities are. On the other hand, marketing personnel are likely to see the manual as a marketing medium, and will expect it to describe in glowing terms why this product is better than ever, what improvements have been made, and why it is preferable to the competitor's product. Service people will focus on the manual as a maintenance and parts resource, and of course, attorneys in the legal department will see it as a

---

* Remember: how it works is not the same as how to use it.

means to reduce the risk of products liability exposure. In fact, the manual, as earlier chapters show, fulfills all these roles and more. Whatever division you are housed in, part of your job is to educate your co-workers to that reality and help to broaden their own perspectives.

Different divisions may have different degrees of power in an organization. The "core" divisions, like engineering and manufacturing, historically have wielded more clout than the "support" divisions, like marketing and service. But that is beginning to change, as mind-sets change about what the product really is. Increasingly, many companies are finding that it's tough to gain a competitive edge by designing a better mousetrap—the designs are all pretty much state-of-the-art. But you may be able to sell more mousetraps if you provide better customer service than the other guys ("Here, let me empty that for you. . . ."). As the balance shifts from the product to the customer, the balance of power also tends to shift within a company. Whatever your placement and whatever the pecking order in your company, the nature of your job requires good relationships with all departments.

CONSENSUS BUILDING AND TURF WARS

As a technical writer, you are in a unique position to assist in bridging the different perspectives of the various departments. More than almost any other job in the company, the job of a technical writer requires you to see the big picture. You are a generalist in a world of specialists. You have to understand the inner workings of the product almost as well as the engineers, you have to understand the needs of the customer as clearly as the marketing professionals, and you need to know how instructions and warnings play into product safety and liability avoidance. In fact, just about the only other person in the company with the need to have such a broad view is the CEO!

With your interdisciplinary perspective, you are in good position to help build consensus among stakeholders and avoid turf battles. You can help the engineers see the need for user-friendly manuals, and you can help marketing see the need to keep product sales literature consistent with the instructions in the manual. You can use your skills as a translator to help all these different departments communicate with each other better and together better serve the customer. The more that you can pull different parts of the company together to work on product and manual development, the better product you will create.

HOW WELL CAN YOU JUGGLE? MANAGING DOCUMENTATION PROJECTS

Like managing any project, documentation requires a vision of the completed project, knowledge of all the components that must come together to create it, an understanding of necessary sequence, and the ability to turn on a dime and go to Plan B when Plan A no longer looks feasible. If you are the manager of a technical publications department, you have an especially challenging position because of the cross-functional relationship building required and because your projects (manuals) may be seen as secondary to the "real" product. To succeed, you need to be able to do all these:

- Give your writers what they need to do the job.
- Schedule projects, assign personnel, and monitor progress.
- Overcome organizational obstacles.
- Advocate effectively for resources to improve your products.

WRITERS NEED ONLY TWO THINGS

Assuming they're reasonably competent to start with, writers need only two things to produce high-quality manuals: information and time. The earlier part of this chapter discussed ways that technical writers could get information. Time is often an even scarcer resource. More errors and slap-dash jobs can be explained by time pressures than by incompetence. Managers often need to be reminded that writing takes time. Writers themselves usually do not have to be convinced.

Technical writers know that writing takes more time than anyone would ever guess, but even writers themselves often underestimate how much time anything takes. You're only putting a bunch of words on a page, right? So, what's the big deal? Consider that an average one- to two-page business letter or memo may sometimes take more than an hour to compose. Now add up the number of pages in a manual. See? The good news is that once writers and/or managers have set up effective, consistent ways to collect information and have gone through the manual production process at least once, subsequent manual production proceeds more quickly.

However, totally new products need especially generous lead times for creating the manual. Some of the vital information may not be available until the last minute, for example, when the prototype is completed and tested. Ideally, you should always build in extra time to track down and talk to all the subject-matter experts involved, and to accommodate those guaranteed-to-appear last-minute changes. Nevertheless, manual writing will always be deadline writing. Writers who have worked in any way on at least one manual already know this. The trick is to make the deadline as realistic as possible. Show this section to your boss. Again. Managers need to be reminded of this fact when they ask the impossible of writers. They'll still ask, but maybe with a bit more sympathy and a few more hours.

SCHEDULING DOCUMENT PRODUCTION

Writing manuals for constantly changing products is difficult enough; scheduling and tracking these writing projects is even worse. Managers of technical publications departments find themselves often faced with multiple projects, shifting deadlines, tight budgets, and limited staff. In such circumstances it is hardly surprising to find many managers operating in constant crisis mode.

Manual-writing projects are difficult to schedule for the same reasons that manuals are difficult to write: you don't have control over information and time. In addition, the manager has to work within a budget that is dictated from above. Further compounding the difficulty is that some managers operate under the misconception (sometimes fostered by writers) that writing is a fundamentally different sort of activity from engineering or manufacturing and therefore cannot be planned, scheduled, and monitored using the same methods. Such managers may assign a writer a project and a deadline, but then have no way of assessing progress—unless the writer comes and says he cannot meet the deadline. By then, of course, it is too late to conveniently add staff to the project and the department is back in crisis mode.

In fact, writing manuals is essentially similar to any design activity, with predictable inputs and outputs that can be set up as a series of milestones or laid out in a Gantt chart. Scheduling and monitoring a documentation project not only permits a manager to make adjustments before a project is way behind, but it also permits building a track record that can help in estimating future projects.

Planning a documentation project requires that you answer three basic questions:

- What do you need to produce?
- How much time do you have to work with?
- What personnel can you use on the project?

These seem obvious, but it is surprising how many managers do not take the time to look systematically for the answers. Let's look at each one.

What Do You Need to Produce?

Before you can begin to do any realistic scheduling, you need to have a very clear idea of the nature of the documentation project itself. Are you writing a user's guide or a service manual? Or both? Are you working solely in paper documentation or will you also be responsible for producing or coordinating one or more videos? Identify as clearly as possible all the types of documents or the multiple purposes of a single document.

When you know what you are going to produce, begin to plan the documents in more detail. Of course, most of the time this planning activity will be a team effort between manager and writers, but whoever does the job, it still needs to be done. This detailed planning stage involves the following kinds of activities:

- Estimating page count for the final document (including front matter and back matter)
- Preparing a detailed outline of the document
- Estimating graphics requirements
- Identifying tasks required to complete the project (writing, editing, interviewing subject-matter experts, preparing graphics, designing page layout, etc.)

When this stage is finished, you should have a pretty clear idea of the size of the project. It is time to try to fit the project to the time available.

How Much Time Do You Have?

Typically, you will be working under a deadline imposed by someone else, such as the shipping date of the product—determined by marketing. Your job will be to work backward from that deadline to the present to find out how much time is available. Here is the procedure:

1. Backtrack from that shipping date however long it will take for prepress activities, printing, binding, and packaging the manual. This is your real deadline.
2. Get out a calendar and count the working days (no weekends or holidays) available between now and then and total them for each month.
3. Multiply the number of days by six to find the hours available if you put just one person on the job (use six instead of eight to permit time for meetings, responding to phone calls, and so forth—and adjust the multiplier as experience dictates).
4. Compare the hours available with your projected page count. From experience, you probably have some idea of how many hours it takes to produce a page of final copy (including first draft, second draft, editing, and graphics). Typically, companies see a range of values, depending on the complexity of the material and the experience level of staff. Anywhere from four to ten hours per page is pretty common.
5. Use the projected page count and hours available to determine how many staff you will need to assign to the project. For example, if your page count is 150 pages, and it takes ten hours to produce a page, you will need to have fifteen hundred hours to do the book. If you only have five hundred hours available, you will need to assign three staff members to work on the project full time in order to meet the deadline.

What Personnel Can You Assign?

Unfortunately, you will seldom have a full set of writers, editors, graphic artists, and assistants available for assignment. They will all be in various stages of working on other projects. What you have to do, of course, is pull somebody off a project that is nearly finished to get the new project rolling and then add others as they become available.

This kind of juggling act is the Tech Pubs manager's principal activity. It may not be quite as stressful as being an air traffic controller, but it shares the element of needing to keep track of a dozen different things at once. The more systematic you can be about assigning tasks and estimating project requirements, the more smoothly the work will flow and the more likely you will be to meet the deadline. To be systematic, you have to have good information about who is doing what and how far along they are—and that is where project monitoring is essential.

MONITORING AN ONGOING PROJECT

Monitoring a project is simply keeping track of how far along the work is and comparing that to the plan, rather like measuring actual expenditures of funds to a budget, and then adjusting for variances. Two tools will help you: benchmarks and project logs.

Benchmarks

Benchmarks are similar to quotas, except they measure progress in terms of time rather than quantity. After you have gone through the manual production cycle a couple of times, you will start to have a sense of how long each stage should take. Exactly what these benchmarks are depend of course on the type of product and complexity of the manuals, but they are likely to be fairly consistent within a single company. Just as a construction project manager can estimate time and costs for architectural and engineering work, preliminary site work, construction, mechanical, and finish tasks, you will become able to estimate time and cost for a writing project. As a simple example, one manufacturer uses the following rules of thumb:

- The first draft will take 60% of total time.
- The second draft will take 20% of total time.
- The remaining 20% will go to editing, project management, coordination with graphics, and other support activities.

You know already how many hours you expect the project to take. If you have used up 30% of the time and have only one-fourth of the first draft done, you probably need to make some adjustments.

Project Logs

Keeping track of the project requires that the writers and others working on a project log their time and activity. Most companies have some sort of weekly time sheet on which employees show how many hours they spent on which project. Normally, these are used to allocate costs among different projects. A time sheet used in conjunction with a brief narrative report of specific activities can give the manager all the information he or she needs to determine whether or not a project is moving on schedule. Of course, the same information (total project hours, hours per page, etc.) can be used for cost estimating: just multiply hours by the salary (plus overhead) for each employee assigned to the project.

The kind of scheduling and tracking that I discuss here does take some time to set up and keep current. There are computer tools to make it easier, ranging from simple spreadsheets to software designed specifically for project management. Whatever aids you use, it still does take time. Is it worth it? The answer to that depends on your situation. If you are a "one-man band," you probably do not need a complex system to track your work. If you are the manager of a 15-person publications department, you probably do need some system to stay on top of progress and recognize the need for adjustments before a crisis develops. If your situation is somewhere in between, you will have to balance the benefits of having the information against the costs of the time it takes to run the system. Remember that the record keeping is not an end in itself—the goal is to produce useful information, not to take up more time than producing the manuals. Don't make it any more complicated than it needs to be.*

---

* As a former co-worker of mine is fond of saying, "You create a monster, you have to feed it."

The sort of scheduling and monitoring system that I have described does have one additional advantage regardless of the size of your operation: it allows you to build a set of data that you can use to make your estimates more accurate for future projects. If you keep track of the hours per page that a user's manual actually requires, after you have produced three or four of them, you should be able to estimate quite accurately, and you will have hard quantitative evidence to use if you need to lobby for more time or more staff. Your boss may not know about writing manuals, but he or she does know about spreadsheets. You will be speaking a language that is understood in the business world—and that will make your request more credible. Implementing a scheduling system may not solve all your problems of short deadlines and multiple projects, but it will probably make meeting those deadlines a little less chaotic.

OVERCOMING ORGANIZATIONAL OBSTACLES

The setting and organizational structure in which a writer operates can be the single most important factor in good manual production. Structures affecting manual writers vary enormously from company to company. These variations are sometimes attributable to the size of the company, to managerial philosophy, or to the maturity of the product. Here are some of the patterns that affect the manual writer's job.

Company Size

The very large company typically has a divisional organization, a diversity of products at scattered geographical locations, and separate cadres of technical writers specializing in manuals for each product category. Further, by the time a company goes national or international, its product line is usually "mature," that is, the product has been around for some time, and the vocabulary for its parts and systems is quite well-established and standardized. (Notable exceptions are the quick-growth electronic and computer technologies and those companies specializing in development of brand-new experimental products.)

Very large companies with a team of technical writers whose sole responsibility is manual publication have the luxury of identifying and selecting good communicators from their own ranks. Alternatively, when they choose to hire new employees from outside, these companies usually have developed interviewing and testing systems to help them select the most qualified applicants. Quite often in large companies, writers come up through the ranks, transferring from parts or service manual writing or from positions in product safety, marketing, or advertising. They bring to the job an in-depth knowledge of the product. Large companies are also able, through their service publication managers and editors, to identify writers who need help with their writing skills. That help is provided by one-on-one editorial assistance, on-the-job orientation, and periodic training sessions.

The publication capabilities of large companies often exceed those found in the formal publishing world. Fully equipped photographic labs; sophisticated computer-controlled presses; computerized systems for layout, format, and translation and dedicated workstations; full-color duplicating machines; and in-house personnel specializing in art and technical drawing, slide production, video and digital imagery—all these are tools of the trade available at many large installations.

The great facilities at a large company may come at a price. Writers may have less autonomy and considerably less flexibility in deciding how best to do their job. If they are at widely scattered locations, they find that information takes longer to travel; filing systems may become harder to tap. If the large company is also decentralized, writing quality may be difficult to control. The manuals produced in Kentucky, for example, may be markedly different in quality and style from those produced in Florida—or France. Because the large company tends to be more rigidly hierarchical, a decision to correct an error or to change the way manuals are done may take years, rather than months, to put into operation. In brief, what is gained through bigness, diversity, and sophistication may be lost through unwieldiness and lack of coordination.

The small and intermediate-size company, on the other hand, typically has one or only a few locations. Such companies tend to be regional and centralized and to have a limited product line. Quite often, the product is young and innovative and consequently there may be no old manuals to use as guides and no well-established vocabulary for parts and systems.

For the writer, the small company can be an exciting and challenging place to work. A young product demands a fresh approach to the manual, and writers can literally create the vocabulary and the approach. Further, writers are less likely to have to deal with inertia or with "we've always done it this way" frustration. Designers and engineers are likely to be more accessible to answer questions, and decision making is usually more fluid and flexible because the small company hierarchy has fewer layers. In fact, some of the most inventive ideas for manual production and layout come from the small companies lucky enough to have creative writers who had to build a first-time manual from the ground up.

Many small and intermediate-size companies assign manual writing to a single individual or to a small group of writers. These writers may be confronted with an awesome array of tasks. They must learn the technology of the product; plan layout; write text and safety messages; arrange for artwork, photos, and drawings; negotiate with printers; edit; and choose paper stock and typefaces. Publication support systems may be spotty in the small company (often little more than a computer and a desk tucked into a corner of an office), and much of the production work must be contracted for. Manual writers who work "solo" feel the pressure of multiple responsibilities and are often rushed and isolated.

Understanding Internal Dynamics

Every company has its own internal peculiarities, its hierarchies, and pecking orders. Writers work within those pecking orders, and situations will inevitably arise in which one person or unit has priority over another. Most writers can live comfortably with lines of authority, if they know what they are. What employees (writers included) find difficult are confused, pass-the-buck procedures in which the lines of authority are never articulated or clearly established and where hidden agendas dominate. As a supervisor, you may know where final authority lies for decision on the manual, but you may neglect to convey that information to your writers. You should try to let your writers know about situations in which their decisions are likely to be superseded by someone with higher authority.

In manual writing, the most common problems with lines of authority arise in the following procedures:

- Determining who will have final say and sign-off on the manual's technical accuracy
- Deciding on appropriate language levels for manuals (engineers and lawyers are often disturbed by the simplicity required for general-public users)
- Deciding on final authority when distinctions must be made between legal safeguards and engineering safeguards
- Deciding on final authority when an editor and a writer disagree sharply on word choice, format, or stylistic preference

Whenever possible, decisions like these should be made by discussion and consensus, with writers included in the discussion. However, when negotiation is clearly not an option, let writers know where final authority lies.

When lines of authority are confusing or fuzzy, it's a good idea to follow up meetings with a memo to everyone in attendance outlining what you understood to be the decisions reached. That way, if you misinterpreted something, others can provide the correction. And if your interpretation is not disputed, you have written evidence that people were given the opportunity.

Writers must remember that Tech Pubs sits in that odd position discussed earlier: you are dependent on input from other departments, but no one (except the customer!) depends on your output. For that reason, it can be frustrating trying to get key people in other departments to act, either to provide information or sign off on manual text or illustrations. After all, your priorities are not their priorities. As suggested, use your networking skills to build relationships. You will be rewarded with cooperation and responsiveness.

HOW TO SELL YOUR BOSS ON HIGH-QUALITY DOCUMENTATION

Good-quality documentation takes resources, and in most business environments, resources are limited and managers compete for them. For Tech Pubs to get more resources, some other department will get less. For business, normally the only reason to spend money is to make money. So, the question becomes, how can you convince your boss that putting more resources into writing manuals will make money? After all, it's not as though you can sell them on the racks at the supermarket between the Westerns and the murder mysteries. Your boss may point out that nobody ever bought a product because they really, really wanted the manual. So how can you show that good manuals make money?

While you cannot easily show that good manuals increase revenue (although they may in fact do so through repeat business from happy customers), you can show that good documentation decreases expenditures. The net result is the same. Good documentation saves money in two important ways:

- Fewer costs for technical assistance or service
- Fewer costs to defend or settle products liability lawsuits—or worse, pay judgments

The first is relatively easy to measure. The second is a little more challenging.

Technical Assistance and Service Costs

Every time someone calls the 800 number for technical assistance, the company is spending money. Someone has to answer the phone and troubleshoot the problem. If there are a lot of calls, the company will need to hire more people and add phone lines to handle the volume without creating unreasonable wait times for already-frustrated callers. Of course, these workers also need to be trained, supervised, and provided with space to work. For companies with lots of products on the market, these costs can mount up fast, particularly when they can't charge the customer directly for the service. Many software companies offer technical assistance contracts to customers as a way to recover their costs, but such contracts are not common with other kinds of products.

Similarly, service calls cost the company money, especially during a product's warranty period (which for large capital machinery may be measured in years rather than days). Very often, the problem may turn out to be operator error rather than a true product defect, but the manufacturer is still out the cost of the service visit. With an on-site service call, the manufacturer has to pay not only the cost of the employee's time during the visit, but also time and travel expenses to get the service employee to the customer's location. Even if a company were to charge for service calls during the warranty period (which might drive away customers), it is unlikely that it would be able to recover the entire cost.

Could better manuals reduce the need for 800-number customer assistance calls and on-site service visits? Absolutely—if the customers can solve the problems themselves (or avoid them in the first place), they won't be calling the manufacturer for help. Remember, the time that the customer spends trying to rectify a problem represents a significant cost to the customer in lost productivity. What you need to do is make the case to your boss. Here's how.

Spend some time getting to know the manager of the service department. Explain that you are doing a research project that involves tracking the time and costs associated with customer assistance hotline calls and service calls. Chances are that the manager will be delighted to talk about this with you; the service department is underutilized as a source of information in most companies. If not, you may need to ask your boss to smooth the way for you. The information you want is this:

- The number of calls/visits for a given model over a period of time, such as the last six months or year. If there is a model that is due for a manual revision soon, choose that one.
- The per-call/visit cost, on average. Don't forget to include associated costs, such as travel expenses.

Calculate the anticipated cost of creating a standard revised manual for the product. Often, scheduled revisions don't really change much in the manual other than updating it about new features. If the manual is in need of more extensive work to make it user-friendly and state-of-the-art, calculate the cost of a substantive revision. The difference between the routine update and the more extensive revision is the critical number. Based on the information you obtained from the service department, you can do a simple cost-benefit analysis.

Here's a simple example:

- Service visits per year (200) × average cost per visit ($1200) − $240,000
- Upgrade revision ($120,000) − standard revision ($80,000) = $40,000

If the upgrade results in 34 fewer service calls (17 %) it will have paid for itself—and that's just for one year. That one-time investment in the manual upgrade will continue to reduce costs for every year that the product is eligible for free service calls—or even paid service calls if the payment does not entirely cover costs.

Depending on the product and the manual, the cost-benefit analysis may show that it would not be cost-effective to upgrade the manual on the basis of service visits or customer assistance calls. But it might still be prudent to do so if safety is at issue.

Products Liability Costs

As Chapter 8 explains in detail, product safety and liability prevention is an issue that increasingly involves instructions and warnings. The cost of ignoring these concerns can be considerable. According to a recent study of 187 products liability lawsuits, the plaintiffs won outright in 40% of the cases, with a verdict of comparative liability (some degree of fault apportioned to both sides) in another 10%. A 2012 insurance industry study found the average jury award in products liability cases to be just under $3.5 million, with the median being $1.5 million.[4] Bear in mind that the overwhelming majority of cases settle before trial, but even those involve paying some amount of money. And the award or settlement amount is only part of the cost of defending against a lawsuit—the cost also includes attorney fees, expert witness fees, research, the time of employees who must prepare and testify, and so on.

Of course, not every injury results in a lawsuit, and not every lawsuit is won by the plaintiff, but it stands to reason that if you can reduce injuries, you will reduce the chances of being sued. While you may not be able to show that putting resources into improving documentation will save you from losing a lawsuit, you may be able to show a reduction in injuries reported following an improvement to the manual. And you can be assured that acting proactively to improve user safety will stand you in good stead even if you are sued. It would be worth your while to do a little research not only on your own company's injury and lawsuit histories, but those of your industry as a whole. You may not be able to calculate the benefit of a manual upgrade out to the third decimal point, but you will be able to cite trends and estimate benefits.

The advantage to conducting this research, even if it is not exact, is that you will be able to quantify the benefits of improved documentation in financial terms. Managers understand budgets and cost-benefit analyses. By presenting your case in terms your manager is comfortable with, you will have a much better chance of getting his or her backing for increased resources.

SUMMARY

Knowing the user and the product is certainly critical to producing good manuals, but so is understanding something about the people who write them and the settings in which they work. Certainly, the profession of technical writing has evolved over time. Technical writers today are better trained, more professional, and better paid than at any time in the past, but they still struggle with many of the same problems they faced decades ago. To produce good documentation, a writer needs accurate and timely information about the product, and getting good information requires the ability to move easily across the organizational lines of the company.

Whether you work alone or as part of a large technical publications department, you'll need to cultivate relationships outside your immediate area. The more people you know and the more you can become involved in product development work, the more likely you will be to have easy access to the information you need to do your job. At the same time, your involvement can help improve the product. You have the dual role of being the voice of the company to the customer and an advocate for the customer to the product development team. Concurrent development, when the product and manual grow together, typically produces the best versions of both.

As a technical publications manager, your task is to take what often seems like a chaotic, creative, wholly unsystematic process (writing a manual) and find a way to schedule it, assign staff to it, and monitor progress. Good record keeping can help you learn from past efforts how better to predict and manage the next project. Good records also enable you to talk to your boss in terms he or she understands, and translate the writer's process accurately into the dollars and time required to do a good job. As the saying goes, ask the boss this: what do you want the manual to be:

- Good?
- Fast?
- Cheap?

Pick any two.

The fact is that high-quality documentation may cost more to produce, but it may well generate an excellent return on investment in the long run. Good manuals add value by encouraging return business and save costs by reducing the need for technical assistance and service calls. Increasingly, good product manuals can significantly reduce a company's liability exposure as well.

CHECKLIST: WRITERS AND DOCUMENTATION PROJECTS

- Have I involved the graphic artist early in my writing process?
- Have I found key contacts in engineering, production, marketing, legal, and other company areas?
- Do I go regularly to product development meetings?
- Have I asked the service department for information about cost and frequency of repairs or technical assistance calls?

- Do I keep good records of how I spend my time so I can accurately estimate the cost to produce a manual?
- Am I monitoring ongoing projects closely enough to recognize when we need to juggle the schedule and reassign personnel?

NOTES

1. U.S. Department of Labor, "Bureau of Labor Statistics, Occupational Employment Statistics," www.bls.gov/ooh/media-and-communication/technical-writers.htm#tab-5, accessed May 23, 2019.
2. Nina Wishbow, "Home Sweet Home: Where Do Technical Communication Departments Belong?" *ACM SIGDOC Journal of Computer Documentation* 23, no. 1 (February 1999): 28–34.
3. Karen A. Schriver, *Dynamics in Document Design* (New York: John Wiley and Sons, Inc., 1997).
4. See *Products Liability Litigation*, Issue 37, Winter, 2015, Smith, Gambrell & Russell, LLP, www.sgrlaw.com/ttl-articles/products-liability-litigation/, accessed May 23, 2019.

11 Integrated Product Safety

The Manual Is Just the Beginning

OVERVIEW

By now you have no doubt realized that manuals and warnings cannot be an add-on afterthought at the end of product development, but rather must be planned for from the beginning. The more organically a product and all its collateral elements can grow together, the more effectively they will work together to give the manufacturer a competitive edge. Marketing professionals know that good "branding" is critical to a successful sales campaign—that is, all company products and communications must present a similar image to the public. That same cohesiveness is essential in product safety as well.

Up to now, we have examined individual aspects of manual and warning design—effective writing strategies, how best to use graphics, designing, and testing warnings, to name a few. Now it's time to put it all together to see how building product safety into every step of the process helps ensure a better product, a better manual, and a satisfied—and safe—end-user.

DESIGNING SAFETY INTO THE PRODUCT

If you ask manufacturing executives what keeps them awake at night, you'll most likely get one of two responses: how to stay competitive in a global economy and how to avoid liability. The two are not totally unrelated. As more and more companies become themselves multinational, or at the least use suppliers in other parts of the world, safety becomes more and more difficult to manage. Other countries may have lower (or no) safety standards for products, and foreign manufacturers may be effectively immune from products liability lawsuits, as they may deny the jurisdiction of U.S. courts. Foreign-manufactured products make up a disproportionate share of product safety recalls. For example, "[I]n 2014, Chinese goods constituted 23% of all goods in the United States under the U.S. Consumer Product Safety Commission's (CPSC) jurisdiction, but represented 51% of all product safety recalls posted by the CPSE."[1] Part of the answer to both maintaining competitiveness and reducing liability lies in taking a comprehensive and integrated approach to product safety. The first step in that direction is to form a product safety team.

Why You Need a Product Safety Team—And Who Should Be on It

A recurring theme in this book has been that safety cannot really be separated from any of the elements that go into the making and selling of a product. Slapping a warning label on a product doesn't fix a design defect. Inappropriate marketing can lead to product misuse and products liability lawsuits. Failure to provide adequate instructions can leave users vulnerable to injury. Yet despite all these connections, many companies leave safety to one individual or one department in the organization. Typically, safety becomes the responsibility of the legal department, the marketing department, or (sometimes) the engineering department.

Why is giving one person the job of managing safety a bad idea? Doesn't it make more sense to vest one individual or one department with such a crucial responsibility rather than spread it out across several? The more a responsibility is spread, after all, the more it tends to become diluted or orphaned with no one taking ownership. In some cases that's true, but safety is an exception. Putting safety in the hands of a team is usually a better choice for three reasons:

- You need multiple perspectives.
- You need cross-functional buy-in.
- You need enforcement authority.

Each of these makes the team approach more effective.

Multiple Perspectives

There's an old saying that if the only tool you have is a hammer, everything starts to look like a nail. To a degree, whatever functional perspective we have becomes our "hammer." If you're an attorney, you tend to look for legal ways to solve problems; if you're an engineer, you tend to try to find engineering solutions. It's perfectly natural for all of us to rely on our own expertise to solve a given problem because it's what we know best and what we do best. But it can also blind us. An engineer looking at a particular product safety issue may search for an engineering solution, because that's the most familiar approach, but what if there isn't one? A different approach is needed, but finding it may require a different person with a different "hammer."

With a team, especially if it is truly a cross-functional team with representation from Engineering, Production, Marketing, Technical Publications, Legal, and Service, you are more likely to find the most cost-effective solution simply because you're covering all the angles. Each member sees the problem through the lens of his or her role and thereby multiplies the chance of finding the best answer that will both improve safety and keep the price competitive.

A cross-functional team often results in more creative solutions as well. We all tend to get stuck in habitual thinking patterns. Bringing together people from multiple disciplines can help spark new ways of looking at a problem. A technical writer who does not have an engineering degree doesn't know how machines "ought to be" designed, and therefore isn't bound by standard practice. Perhaps there is an engineering solution after all—just not a traditional one.

Cross-Functional Buy-In

For any new approach to succeed, you must have buy-in from all parts of the company. If you don't, the new initiative will be actively sabotaged, passively resisted, or simply ignored. People generally don't like change, even when the change is good—unless they see it as their own initiative. Creating a safety team with members from all parts of the organization means not only that all the stakeholders are represented at the table, but also that whatever the team decides is to some degree "owned" by each part of the company. Because the whole company needs to be involved for product safety to be truly integrated, a cross-functional product safety team is essential.

Enforcement Authority

We've all been part of teams and committees that accomplished nothing except eating up time in pointless meetings and creating reports that no one reads. Usually when that happens, one of two causes can be found:

- The committee was formed to give the appearance of addressing a concern that no one really intended to fix.
- The committee had no authority to make decisions.

The first of these is usually not a problem with a product safety committee. If product safety concerns are keeping your CEO awake at night, he or she most likely in fact does want to fix the problems. The second flaw is all too common. If all the members of the committee are relatively near the bottom of the organizational chart, their recommendations are likely to go unheeded because they do not carry the weight of the real decision makers in the organization. On the other hand, if all the members of the team are department heads, the team will not have the perspective of those "in the trenches," and will not be likely to appreciate all the factors involved to make a safety solution workable.

Even more important than membership, however, is commitment from the top of the organization. Company leaders need to send a clear message that they take safety seriously, that the product safety committee is not just "for show," and that the leadership will back its recommendations. Without that clear message, product safety efforts will be an uphill struggle at best. If the commitment is there, however, the product safety team can make a real difference not only by improving safety, but by improving product design, documentation, and marketing at every stage of development. This process is called "concurrent development."

CONCURRENT DEVELOPMENT: BETTER AND SAFER PRODUCTS

The idea of concurrent development is simple: at the same time that the product is being developed from drawings to prototype to production model, all the associated documentation, marketing, and safety materials are being developed as well. These development processes operate in parallel rather than serially (one after the other). One obvious benefit of concurrent development is that the finished product does not sit in a warehouse waiting for the manual to be written before it can be shipped out to paying customers. Another benefit is that concurrent development can actually improve safety.

As discussed in Chapter 2, the hierarchy for managing hazards is to design them out if possible, shield or guard them if they can't be designed out, and warn against them (either with an on-product label or in the manual) if they cannot be guarded. If you wait until the product is already developed before conducting a hazard analysis, chances are that it will be too late to find design solutions to any of the safety issues. The result is that more warnings will probably be needed—and that means that the impact of any one warning will be lessened. If an injury results from one of the product hazards, the manufacturer may be found liable and may have to pay a significant judgment. Taking an integrated approach to product safety prevents many of these problems. Let's look at an example.

Suppose your company wants to manufacture an industrial wet-vacuum. Your product idea is a motor-and-hose assembly that mounts on a 55-gallon drum provided by the customer. You will also provide a wheeled base to mount the drum for easy movement. While the product is still in the design-sketch phase, you bring together your product safety team and conduct a hazard analysis. You identify six potential hazards:

- Slip-and-fall injuries to workers operating on wet floors
- Electrocution/electric shock hazard if the appliance is not grounded (the unit design features an attached three-prong plug for use with the customer's extension cord)
- Motor damage resulting from voltage drop if the product is used with an extension cord that is too long or not of adequate gauge
- Fire or explosion hazard if the unit is used to clean up flammable liquid spills
- Tip-over hazard if the product is not mounted on the wheeled base provided
- Toxic exposure if the unit is used to clean up hazardous material or if the drum previously contained hazardous material

In analyzing these hazards, the product safety team makes the following decisions:

- The slip-and-fall hazard is open and obvious—if the wet-vacuum is being used to clean up a spill, the user knows that the floor is wet, and wet floors are generally known to be slippery. Open and obvious hazards ordinarily do not need warnings. The hazard could be addressed in the accompanying instructions.
- The electrocution/electric shock hazard should be addressed in a label on the product. The likelihood of this hazard occurring is high (people are known to use adaptors or to cut off the grounding plug) and the consequence is severe. Additionally, a partial design solution to the hazard might be to provide an attached power cord of sufficient length that the need for an extension cord is reduced.
- The voltage drop could be addressed in the instructions, since the consequence of the hazard is damage to the motor rather than personal injury.
- The tip-over hazard can be designed out, by providing the drum and attaching it to the wheeled base. Doing so would also eliminate the potential of the customer using a drum contaminated with hazardous or toxic material.

- The fire/explosion hazard and toxic exposure hazard could be combined in a single label warning against using the wet-vacuum to pick up flammable or hazardous materials. While the likelihood of this hazard occurring is probably less than the unit's being plugged into a non-grounded receptacle, the consequence is severe.

Without the hazard analysis, the design would be less safe,* the technical writer might not recognize some of the hazards (e.g., the voltage-drop problem), and the product would need to have more warning labels.

In the example just described, while there is an increase in the cost of manufacture (adding a power cord and furnishing the drum), that cost may well be offset in the short run by increased sales because the new design is more convenient (the buyer doesn't need to furnish a drum and a cord), and more attractive (fewer labels). It will almost certainly be offset in the long run by reduced injury and liability claims.

Establishing a product safety team with enough clout to make a difference and following the practice of concurrent development will help ensure that your products are safe and user-friendly when they go out the door. But integrated product safety requires more than simply a good design. You must make sure that the focus on safety extends beyond just design and fabrication.

DELIVERING A CONSISTENT SAFETY MESSAGE

Other parts of this book have alluded to the importance of making sure that all the aspects of product development, marketing, sales, and service work together. While working together is important in many aspects of selling products to the marketplace, it is absolutely essential in matters of safety. Nothing will make a plaintiff's attorney happier than to discover an inconsistency in safety-related information.

WHY THE MODEL WORE SAFETY GLASSES

The inconsistency that makes the plaintiff's attorney jump for joy need not be a verbal contradiction. For example, in less than three minutes, I just found the following on the Internet: a website for a very well-known manufacturer of power tools featuring a picture of a woman using a reciprocating saw to trim plastic-covered wire shelving to size. She is not wearing safety glasses. However, a look at the manual for that same saw (also available on the website) reveals the following statement on page 2:

- Use safety equipment. Always wear eye protection.

And this statement on page 7:

> **WARNING:** Always wear proper eye and respiratory protection while operating this product.

---

* Less safe does not necessarily mean that the design would be defective. It would mean that the company would be more likely to have the opportunity to go to court and find out.

Which is more compelling to the user? A photo of a woman using the saw or a warning embedded in the midst of dozens of other warnings and safety instructions in a manual for a rather simple product? Certainly, the plaintiff's attorney will argue—and provide experts to explain—that visual images are much more likely to be seen and remembered than verbal information, just as body language and tone of voice carry more weight than the actual words we speak.

How about this? A manufacturer of a grain auger puts a warning label on the product near the rotating auger that reads:

WARNING

• Entanglement hazard.
• Keep hands and feet away.
• Disconnect power before servicing or attempting to clear a jam.

The label complies with ANSI Z535.4, and includes a pictorial of a hand becoming caught in an auger. In the accompanying manual, however, the following text appears in the section that describes how to clear a jam:

> **DANGER!** Do not attempt to clear a jam while the auger is running. Your hand and arm could get caught. Always turn off the power to the auger before cleaning or removing stuck material.

The use of two different signal words for the same hazard is certainly inconsistent—and it is doubly a problem, because the on-product label (using the lesser signal word) is much more likely to be seen by the worker than is the manual. A farmer may own the grain auger, but a hired hand may be operating it.

How can these inconsistencies happen, and what can you do to prevent it? The most common reason for a lack of consistency is simply that different people prepare different parts of the product literature. The marketing team is not thinking about safety when they shoot the photo of the woman using the reciprocating saw—they're thinking about making the product look handy to have around the house and not requiring a linebacker's build to operate. The model is certainly not likely to think of safety glasses. In the case of the grain auger, most likely the technical publications department was responsible for the manual and the legal or marketing department was responsible for the labels—and they didn't talk to one another.

Prevention is simple—but not always easy. The problem occurs because the process of bringing a product to market is divided up into separate functional areas; the solution is to ensure that some person or group can see the big picture and check across all the areas for inconsistencies. This role can be played by an individual or a team, but it must be done. Whether you have a product safety czar or a product safety team, the job is the same: make sure that the message you deliver is consistent from one aspect of the product to the next. An important, but sometimes overlooked, piece of the process is the sales force in the field.

THE PATENT MEDICINE SHOW: WHEN THE SALES REP PROMISES THE MOON

Before the days of the Federal Food and Drug Administration, purveyors of patent medicines would travel from town to town, providing free entertainment to entice customers to listen to their sales pitches. They were often remarkably successful, in part no doubt because of their claims that a single elixir could cure a multitude of conditions. While we have moved past the appeal of snake-oil salesmen, we humans are still susceptible to the lure of a product that promises to perform many functions. Consider the smartphone. Not only can you use it to make telephone calls, you can also take pictures, get turn-by-turn driving directions, surf the Internet, play games, read your email, and watch TV shows—among other things.

Combine that human desire for a product to do everything with a sales representative's desire to earn a commission, and the potential for trouble looms large. Even if the sales rep doesn't deliberately overstate the capabilities of the product, he or she may be reluctant to say no when a customer asks if the product can be used in a non-standard application. It's critical that the sales department understand not just the features that make the product attractive, but also the potential problems that can occur with misuse—remember, misuse means only a departure from the instructions for safe use, not necessarily deliberate abuse. One way to get the sales force on board with ensuring product safety is to make sure that they are represented on the product safety team. In turn, the product safety team will benefit because the sales reps spend most of their time in the field. They can provide valuable insight into the product's use environment.

INSTALLATION: THE CUSTOMER CONNECTION

Installation of products is often a critical step in ensuring that the product operates safely. For consumer products, installation may be done by the end-user or by installers working for the seller—installers who may or may not be factory-trained. For large capital machinery, the installers may work directly for the manufacturer or may be the maintenance staff employed by the customer. For consumer products, the manufacturer may have little control over whether the product is properly installed, apart from preparing the installation instructions. When installers are factory-trained, they must clearly understand how important it is to install all safety equipment that comes with the product. They should not leave it up to the end-user, because more often than not, the end-user won't bother.

For industrial products, the connection is much closer. Not only is installation often performed by the manufacturer's own employees, but in many cases the machinery is customized for the buyer. In these cases, usually there is a good deal of communication between manufacturer and purchaser. With custom installations, especially in large operations, the new machine or other product may have to be integrated into an existing system. Installers must make sure that the systems are compatible. For example, electrical systems can be set up to be normally closed or normally open. If the circuit is normally closed, then under ordinary circumstances current is flowing—a fault condition will break the circuit and stop the flow of electricity. If a product is to be wired into an existing system, it will be important to know whether the relays or other switches operating automatic safety devices are designed for a normally closed or normally open set up. The product itself might work either way, but the safety devices might not.

SERVICE—AND SAFETY—WITH A SMILE

As noted elsewhere, service reps are often a manufacturer's best source of information about where instructions and warnings are inadequate. They can also be a valuable resource for delivering safety information to customers. Suppose, for example, that you discover a potential misuse of the product because of an injury that occurs. After studying the incident, the product safety team decides that the hazard needs to be addressed with an additional warning label. How can you get these to your customers? Certainly, the simplest method is just to mail it—but can you be sure that the customer will actually apply the label and not just toss it in a desk drawer? Another approach would be to mail the new label out with a letter, but also give a supply of the labels to the service reps. The service reps can then check on their regular visits to be sure the label is in place, and if not, put one on. Be sure that the service reps document delivery of safety information, application of labels, retrofitting of any safety equipment, etc.

Service reps can offer all kinds of safety services for customers, including providing technical bulletins to address specific safety issues, replacing worn-off labels, or simply giving advice on the best way to solve a problem without compromising safety. As the company's eyes and ears in the field, service reps can be highly effective as a two-way conduit for safety information. Because they are out speaking directly with customers, they establish a person-to-person relationship. That kind of close connection makes it more likely that the customer will pass along information about safety-related issues, including where safe operating procedures are compromised and why. By the same token, that personal relationship makes it more likely that customers will heed the service rep's advice about the importance of working safely. Be sure that the service reps are in the loop with the rest of the product safety team so that the message is consistent from manual to installer to service rep.

INTEGRATED PRODUCT SAFETY IN THE REAL WORLD: CASE STUDY

This book recommends concurrent development and a team approach to address product safety issues. It sounds good on paper, but is this approach really practical in the real world? It is certainly practical, although never as tidy a process as it looks on paper. To illustrate a real process, this section provides a case study in how a safety-conscious and proactive company worked to ensure that its users would be properly protected with good instructions and warnings and that those instructions and warnings would meet the appropriate standards.

The product, the uGO™ FlameDisk™,[2] consists of a solid, ethanol-based fuel intended for use as an alternative to charcoal briquettes for outdoor cooking. It is packaged in an aluminum pan with a perforated aluminum disk crimped and sealed on top.

The holes in the disk allow the fuel to be ignited and burn in a controlled manner. A peel-off film is applied to the top of the disk to seal the container until ready for use.

The user peels off the film, sets the pan on a grate in a suitable grill, and ignites the fuel. The fuel burns for approximately 45 minutes, at a heat level similar to that of charcoal. When finished cooking, the user lets the remainder of the fuel burn off, causing the unit to self-extinguish. After a few minutes, the container is cool and can be disposed of in a normal trash or recycling container.

The process from idea to finished product followed a path typical for many products. It can be divided into three phases:

1. Initial information gathering and hazard analysis
2. Drafting instructions and warnings
3. Refining the drafts and ensuring consistency

Here's a closer look at each.

INITIAL INFORMATION GATHERING AND HAZARD ANALYSIS

In July 2007, I was contacted by one of the owners of Sologear, Inc., the manufacturer of the FlameDisk for product safety consulting services. He provided a video showing the prototype product in use to cook hamburgers, some photos, an engineering drawing, and an oral description of how to use the product.[3] One of the reasons to use a consultant, particularly with a new product, is to help generate "what if" questions. After an initial review of the material provided, I asked these questions of the manufacturer:

1. What happens if it is stored above 120° F? (I'm thinking about Phoenix, Arizona, consumers, for example, where the outside temperatures can easily reach 115—and even as high as 121. If I have some uGO packages stored in my garage, which is not air-conditioned, am I asking for trouble?
2. What if I'm cooking and a piece of food falls through the grill onto the uGO? Is it safe to eat after I pick it up (burning my fingers, of course)?
3. What's the shelf life? Is the product stable at normal temperatures for long periods of time? Is it still safe to use if it's been stored a long time?
4. How should I extinguish it if I have to, for some emergency?
5. Is the product toxic? Ethanol is not (generally), but I assume you've added something to it to make it a solid at normal temperatures.
6. Are there any common substances this product would react negatively with?
7. Are there any things I can do to make the product burn much more rapidly than intended? For example, if I douse it with charcoal lighter fluid before lighting it, would that be dangerous?
8. What would happen if I started to use charcoal and then got frustrated because it was taking so long to get going, so I dumped a uGO in on top?

After further investigation, we identified the following as the primary hazards:

- Burns from contact with burning fuel (e.g., from picking up food that had fallen through the grate or because the user could not see the flame)

- Burns from contact with hot pan containing the fuel (e.g., trying to reposition it while fuel was burning)
- Ignition of nearby combustibles (e.g., if the FlameDisk were set on a picnic table instead of in a conventional grill)
- Excessive flame height and/or rapid burning of fuel if perforated lid were removed
- Ignition of an unopened FlameDisk container if stored on a hot surface or placed in contact with a flame (the product is stable at expected ambient temperatures, even in Phoenix)
- Ingestion of the ethanol-based fuel, either accidentally or intentionally to get "high"

We also determined that as a consumer product containing a flammable substance, the FlameDisk would need to meet the labeling requirements of the *Federal Hazardous Substances Act*.[4] We investigated whether it would need to meet the requirements of the *Poison Prevention Packaging Act*[5] and concluded that it would not.

DRAFTING OF INSTRUCTIONS AND WARNINGS

The manufacturer had prepared a rough draft of instructions, which we amended to the version shown in Example 11.1. As you can see, all of the identified hazards are addressed in the instructions, either by the instructions themselves ("Allow it to fully burn out and cool before handling") or with embedded warnings ("WARNING: Flame may be hard to see. Assume fuel is lit even if you do not see flames.")

Example 11.1: Intermediate Draft of Instructions for FlameDisk™

Use Instructions

WARNING! The uGO must be used in a conventional grill. Never place uGO on a flammable surface. Do not place food directly on uGO.

1. Open all grill vents fully.
2. Remove food grate from grill and set aside.
3. Remove uGO from box and place on charcoal grate in bottom of grill. Note: if using an open flat-pan type grill, place uGO directly on the bottom of grill. [Include two graphics, one showing each application.]
4. Peel off cover and discard, exposing perforated lid. Do not remove lid. CAUTION! Do not touch fuel. In case of skin contact, wash skin thoroughly. If food comes in direct contact with unburned fuel, discard it.
5. Using a lighter or match, light fuel through hole in lid. Light only one hole— flame will automatically spread. WARNING! Flame may be hard to see. Assume fuel is lit, even if you do not see flames.
6. Replace food grate and cook food. For best performance, leave grill cover open. Fuel will cook at about the same rate and temperature as charcoal. If fuel goes out, simply relight. When finished, allow fuel to burn out.

Disposal Instructions

1. Important: Allow fuel disk to burn out completely and let unit cool before handling. Fuel disk will burn for about 45 minutes before fuel is used up. Allow another 10 minutes for unit to cool.
2. Do not attempt to extinguish flames by blowing on them or using water or other liquid. If necessary to extinguish before fuel is used up, close grill cover and vents or use a dry chemical extinguisher rated for Class B fires.
3. When fuel is gone and unit is cool, dispose of properly. Unit and carton are recyclable.

Language was also developed to fulfill the *Federal Hazardous Substances Act* requirements:

> **WARNING:** FLAMMABLE MIXTURE. Contains ethanol and less than 4% methanol. Do not consume fuel. If swallowed, drink 2 glasses of water and call physician or poison control center. In case of eye contact, flush with water and call physician. Keep out of reach of children. Store away from flame, sparks, and heat. Use outdoors only.

In addition to the words-only FHSA warning, we decided to provide an ANSI-style warning label about the primary hazard of flammability and potential for burns. Note that while it may seem open and obvious that the product gets hot when ignited (that's its purpose, after all), the ethanol burns very cleanly with a flame that can be near-invisible. It's possible that someone could look at the FlameDisk and not immediately realize that it was lit.

The draft label contained this language:

WARNING: Fire and Burn Hazard

- Container becomes hot during use.
- Keep away from combustible materials.
- Allow to cool completely before handling.
- Never leave unattended.
- Read instructions before use.

REFINING THE DRAFTS AND ENSURING CONSISTENCY

At the same time as draft instructions and warnings were under development, the final formulation for the fuel was being developed and the packaging for the product was being designed. To ensure that the user would see the instructions and the warning label, it was decided to print them right on the film cover rather than on the outer box. If they had appeared on the box only, it would have been possible for a user to open the box, take out the FlameDisk, and discard the box without seeing or reading the instructions. However, the warning required by the FHSA must appear on the packaging that will be visible when the product is displayed for retail purchase—which means the outer box. Ultimately, it was decided to print the FHSA label twice—once on the outer box and once on the film cover.

After further product development, including naming the perforated disk (Smokerplate™), the instructions took their final form, as shown in Example 11.2.

Example 11.2: Final FlameDisk™ Instructions

Usage Instructions

WARNING! The uGO™ FlameDisk™ must be used in a conventional grill. Never place uGO™ FlameDisk™ on a combustible surface. Do not place food directly on disk.

1. If cooking on a small grill, use one FlameDisk™. For larger grills, use two.
2. Make sure all grill vents are fully open. Remove food grate and set aside.
3. Remove FlameDisk™ from box and place on charcoal grate in bottom of grill. If using an open, flat grill, place disk directly on bottom of the grill pan.
4. Peel off film cover, exposing perforated Smokerplate™. Do not remove perforated Smokerplate™. CAUTION: Do not touch or consume fuel. In case of skin contact, wash skin thoroughly. Discard any food that comes in direct contact with fuel. Avoid breathing fuel vapors.
5. With lighter or match, light fuel through one of the outer holes on the Smokerplate™—flame will automatically spread. WARNING: Flame may be hard to see. Assume fuel is lit, even if you do not see flames.
6. Carefully replace food grate, wait four minutes and cook food. Cooks at about the same rate and temperature as charcoal.
7. For best performance, do not use grill lid. If fuel goes out prematurely, carefully relight.
8. FlameDisk™ will burn for about 40 minutes. Allow it to fully burn out and cool before handling.

Disposal Instructions

1. Important: FlameDisk™ will burn for approximately 40 minutes before fuel is used up. Allow it to burn out completely and then let cool an additional 10 minutes before handling.
2. Do not attempt to extinguish flames by blowing on them. If necessary to extinguish before fuel is used up, close grill cover and all vents, or douse with water, or use a dry chemical extinguisher rated for Class B fires.
3. When fuel is gone and FlameDisk™ is cool, dispose of properly. FlameDisk™ and carton are recyclable.
4. Dispose of unused fuel by inserting FlameDisk™ into a grill and burning, or as required by local regulation.

Figure 11.1 Film cover layout for FlameDisk. Note the words-only warning required by the ***Federal Hazardous Substances Act*** as well as the ANSI-style warning including a pictorial. Instructions printed on this cover ensure that the user will have access to the instructions even if the outer carton has been discarded.

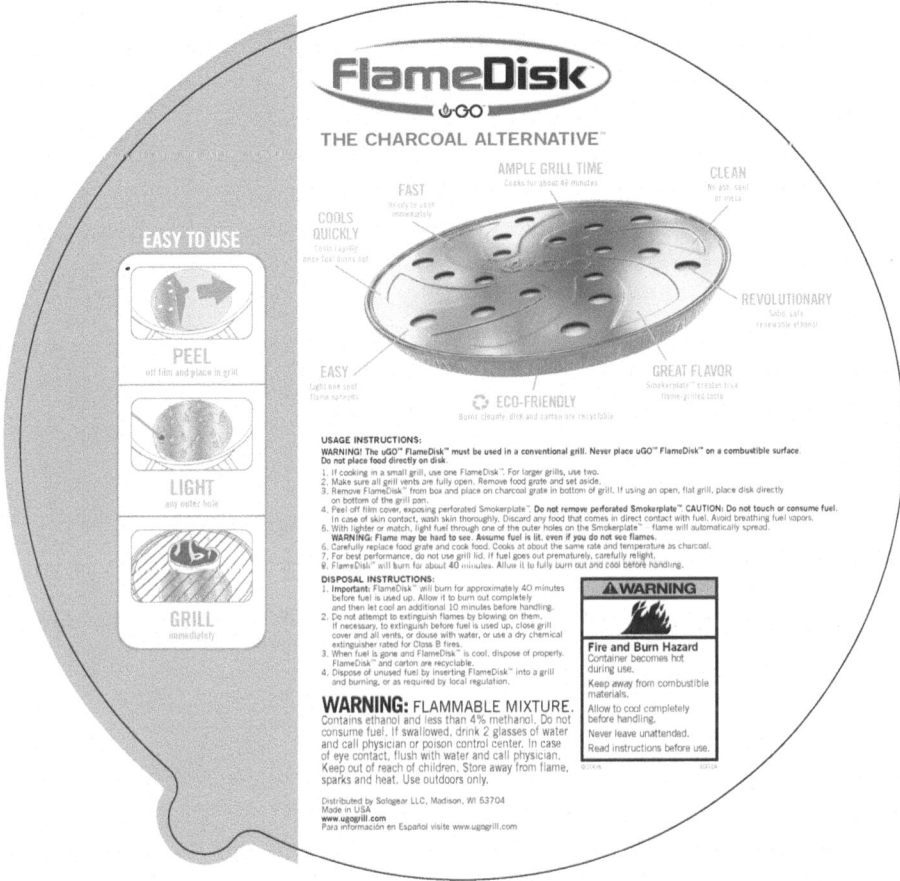

FIGURE 11.1   shows the film cover layout, including the instructions, the ANSI-style warning label, and the FHSA label.

In this case, the manufacturer decided to go one step further to protect the user and provide a third "layer" of safety information. When the user removes the film cover preparatory to lighting the FlameDisk, he or she will see a 2½ × 3-inch "sticky note" stuck directly to the perforated Smokerplate warning that the product becomes hot when lit and that the flame is hard to see (see Figure 11.2). It also directs the user to read the instructions before using the product.

Does the "sticky note" add to the manufacturing cost? It surely adds a little to the cost, but it provides an even greater benefit in ensuring that users will be well-informed about potential hazards so they can use the product safely. This company put protecting the user first. Had they been concerned only with avoiding liability, the "sticky note" could probably have been omitted—the warnings on the box and film cover would almost certainly have been sufficient. But adding the "sticky note"

Warning!

Hot when lit.

Flame may be difficult to see.

Assume fuel is lit even if you do not see flames.

Remove this sheet and light outer hole.

FIGURE 11.2 "Sticky note" warning placed on Smokerplate. This provides an additional layer of safety, conveying essential hazard information and abbreviated instructions even after the film cover has been removed.

ensures that the user will see the most critical warning, even if he or she is bound and determined not to read labels!

The whole process, from first contact to final versions of instructions and warnings, took about six months. As is typical, both instructions and warnings went through multiple drafts (more than shown here), with much discussion back and forth about alternative wording. What made this project easy from the consultant's point of view was the manufacturer's attitude. Instead of resisting the need for warnings, the company recognized that proper warnings actually make the product more user-friendly by ensuring that customers have the information they need to keep themselves safe. This project exemplifies the ideal of integrated product safety.

SUMMARY

Making safety an essential part of the product beginning when the product design is just a gleam in an engineer's eye to the time the manufactured product reaches the end-user (and after) is the most effective way to ensure that your products are both competitive and safe. Integrating product safety into every stage and aspect of getting a product developed and out the door usually requires establishing a product safety team with overall authority to make decisions regarding safety-related matters. The most effective teams include representation from all parts of the company and from multiple levels—those at the top and those "in the trenches." Such teams bring a diversity of perspective and expertise to problems and develop solutions that are more likely to achieve support throughout the company.

Taking an integrated approach to product safety also helps ensure that you deliver a consistent safety message, not only in the instructions and warnings, but throughout the sales and service efforts as well. By looking at all aspects, the product safety team—if it is supported from the top—can make sure that marketing brochures portray the product and its use environment in a way that supports the safety messages in the manual. It can ensure that the sales force represents the product properly, and encourages users to follow safe practices. The team can set guidelines for installers and service reps to focus their attention on making sure that safety issues are properly addressed.

In short, integrated product safety enhances every aspect of production and sales, but most especially, an integrated approach helps those charged with developing manuals and warnings. Building safety into every product from the start ensures that the manual is seen as part and parcel of the product itself, not as a necessary (and expensive) add-on. Concurrent development of product and manual creates better and safer products that can compete in the global marketplace.

CHECKLIST: INTEGRATING PRODUCT SAFETY

- Does my company have a product safety team?
- Does the team have representatives from all parts of the company?
- Is the team empowered to make decisions relating to safety?
- Is all product literature consistent in the safety messages it delivers?
- Has the sales force been briefed on safety issues?
- Have installers been trained to address safety-related installation issues?
- Are service representatives trained to be alert to potential safety issues in the field?

NOTES

1. Matt Snyder and Bart Carfagno, "Chinese Product Safety: A Persistent Challenge fo U.S. Regulators and Importers," Staff Research Report, U.S.–China Economic and Security Review Commission, March 23, 2017, www.uscc.gov/sites/default/files/Research/Chinese%20Product%20Safety.pdf, accessed July 2, 2019.
2. uGO and FlameDisk were trademarks of Sologear, LLC, Madison, Wisconsin. The company was purchased by Bic in 2011, and Bic subsequently manufactured and marketed the FlameDisk charcoal alternative for several years. It was recently discontinued by the manufacturer, but may still be purchased on line at Amazon.com and other online retailers.

3. The process also included normal business documents, including a proposal, confidentiality agreement, etc. Information about the process is presented here by permission of Sologear, LLC.
4. 15 *U.S. Code* 1261.
5. 15 *U.S. Code* 1471–1476. See also 16 *Code of Federal Regulations,* Part 1700 for specific regulations.

Resources

ANSI Z535.4–2011 (R2017). 2017. *American National Standard: Product Safety Signs and Labels.* Rosslyn, VA: National Electrical Manufacturers Association.

ANSI Z535.6–2011. 2011. *For Product Safety Information in Product Manuals, Instructions, and Other Collateral Materials.* Rosslyn, VA: National Electrical Manufacturers Association.

Coe, M. 1996. *Human Factors for Technical Communicators.* New York: John Wiley & Sons, Inc.

Coe, M. 1997. Writing for other cultures: ten problem areas. *Intercom,* 44:1, pp. 17–19.

Cooper, A. 1999. *The Inmates Are Running the Asylum: Why High-Tech Products Drive Us Crazy and How to Restore the Sanity.* Indianapolis, IN: SAMS, a division of Macmillan Computer Publishing.

De Lange, R. W., Esterhuizen, H. I., and D. Beatty. 1993. Performance differences between Times and Helvetica in a reading task. *Electronic Publishing,* 6:3, pp. 241–248.

Dobb, F. P. 1996, 1998. *ISO 9000 Quality Registration.* Oxford: Butterworth-Heinemann.

Ganier, F. 2004. Factors affecting the processing of procedural instructions: implications for document design. *IEEE Transactions on Professional Communication,* 47:1, pp. 15–26.

Globally Harmonized System of Classification and Labeling of Chemicals (GHS). 2007. Second Revised Edition. New York and Geneva: United Nations.

Hofmann, P. 2007. Localising and internationalising graphics and visual information. *IEEE Transactions on Professional Communication,* 50:2, pp. 91–92.

Laughery, K. R., Sr., Wogalter, M. S., and S. L. Young, eds. 1994. *Human Factors Perspectives on Warnings.* Santa Monica, CA: Human Factors and Ergonomics Society.

Lengler, R. 2006. Identifying the competencies of 'visual literacy'—a prerequisite for knowledge visualization. *Proceedings of the 2006 Information Visualization Conference of IEEE Computer Society.*

Lippincott, G. 2004. Gray matter: where are the technical communicators in research and design for aging audiences? *IEEE Transactions on Professional Communication,* 47:3, pp. 157–170.

Melton, J. H. Jr. June 2008. Lost in translation: professional communication competencies in global training contexts. *IEEE Transactions on Professional Communication,* 51:2, pp. 198–214.

Merkt, W., et al. 2011. Learning with videos vs. learning with print: the role of interactive features. *Learning and Instruction,* 21: 687–704.

Norman, D. A. 1990. *The Design of Everyday Things.* New York: Doubleday/Currency.

Ostrom, L. T. and C. A. Wilhelmsen. 2012. *Risk Assessment: Tools, Techniques, and Their Applications.* Hoboken, NJ: John Wiley and Sons.

Peters, G. A. and B. J. Peters. 1999. *Warnings, Instructions, and Technical Communications.* Tucson, AZ: Lawyers and Judges Publishing Co., Inc.

Restatement of the Law Third, Torts: Products Liability. 1998. Philadelphia, PA: American Law Institute Publishers.

Schriver, K. A. 1997. *Dynamics in Document Design.* New York: John Wiley & Sons.

Tufte, E. R. 1983. *The Visual Display of Quantitative Information.* Cheshire, CT: Graphics Press.

Tufte, E. R. 1997. *Visual Explanations.* Cheshire, CT: Graphics Press.

Van Horen, F. M., Jansen, C., Maes, A., and L. G. M. Noordman. 2001. Manuals for the elderly: which information cannot be missed? *Journal of Technical Writing Communication*, 31, pp. 415–431.

Vaubel, K. P. and J. W. Brelsford. 1991. Product evaluation and injury assessments as related to preferences for explicitness of warnings. *Proceedings of the 1991 Human Factors and Ergonomics Society Annual Meeting*, pp. 1048–1052.

Wishbow, N. February 1999. Home sweet home: where do technical communication departments belong? *ACM SIGDOC Journal of Computer Documentation*, 23:1, pp. 28–34.

Wogalter, M. S., ed. 2006. *Handbook of Warnings*. Mahwah, NJ: Lawrence Erlbaum Associates, Inc.

Wogalter, M. S., DeJoy, D. M., and K. R. Laughery. 1999. *Warnings and Risk Communication*. Philadephia, PA: Taylor & Francis, Inc.

Wogalter, M. S., Young, S. L., and K. R. Laughery, Sr., eds. 2001. *Human Factors Perspectives on Warnings, Volume 2*. Santa Monica, CA: Human Factors and Ergonomics Society

Index